Origins and Extinctions

CONTRIBUTORS

Alan H. Guth is professor of physics at the Massachusetts Institute of Technology.

Lynn Margulis is University Professor in the Department of Biology at Boston University.

Donald E. Osterbrock is professor of astronomy and astrophysics at the University of California, Santa Cruz.

David M. Raup is chairman of the Department of the Geophysical Sciences and Avery Distinguished Professor at the University of Chicago.

Peter H. Raven is director of the Missouri Botanical Garden in St. Louis.

George W. Wetherill is director of the Department of Terrestrial Magnetism of the Carnegie Institution of Washington.

Origins and Extinctions

EDITED BY

DONALD E. OSTERBROCK AND

PETER H. RAVEN

*Based on a Symposium on Life and the Universe,
held at the National Academy of Sciences,
Washington, D.C., April 30, 1986*

Yale University Press
New Haven and London

Designed by James J. Johnson
and set in Aster Roman type by
Keystone Typesetting, Inc., Orwigsburg, Pa.
Printed in the United States of America by
Murray Printing Company, Westford, Mass.

Library of Congress Cataloging-in-Publication Data

Symposium on Life and the Universe (1986 : National Academy of
  Sciences)
    Origins and extinctions : based on a Symposium on Life and the
  Universe, held at the National Academy of Sciences, Washington,
  April 30, 1986 / edited by Donald E. Osterbrock and Peter H. Raven.
  p. cm.
  Includes bibliographies.
  Contents: Foreword / Donald E. Osterbrock and Peter H. Raven —
The birth of the cosmos / Alan H. Guth — Formation of the
Earth / George W. Wetherill — The ancient microcosm of planet Earth
/ Lynn Margulis — Extinction in the geologic past / David M. Raup.
  ISBN 0–300–04260–4 (cloth)
      0–300–05471–8 (pbk.)

  1. Earth—Origin—Congresses. 2. Cosmology—Congresses.
I. Osterbrock, Donald E. II. Raven, Peter H. III. Title.
QB631.S95 1986
523.1'2—dc19   *88–1396*
                                                    CIP

10 9 8 7 6 5 4 3 2

To

Bryce Crawford, Jr.

Former Home Secretary of the National Academy of Sciences

Contents

Foreword

What subjects can possibly be more interesting than life and the universe we live in? All civilizations, all peoples we know of have been interested in the universe—what it is, how it started (if it ever did), how it will end (if it ever will).

Who are we? Where did we come from? Where are we going? On a cosmic scale, these are astronomy's big questions. Every ancient clan, every tribe, every nation, every religion has had its own answers. Our ancestors inherited their answers in the Bible. Copernicus provided some new answers. So did Galileo. Every scientist since has been looking, in one way or another, for further answers to these universal questions.

Since Edwin Hubble's pioneering observations in the 1920s, we have known that the universe is expanding. Can we trace it back to its beginning in time? Or in space? Or in both? Or in neither? Are these distinctions relevant? Or meaningful? How could the inhomogeneities, which ultimately became galaxies, form? How did stars form within them? How did planets form around the stars? How did life start on these planets? We cannot answer all these questions, but we can try to provide partial answers to some of them.

The first question, tracing the universe back to its beginnings, or as close as we can get, is intimately tied up with the constitution of matter. It depends on the equation of state at ultrahigh densities, which is related in turn to the highest-energy particle interactions.

Theoretical concepts of "inflation" and "strings" are important parts of this subject, the study of "grand unified theories," known to the experts as GUTS.

Our expert in this field is Alan H. Guth of the Massachusetts Institute of Technology, who is the creator and developer of many of these ideas. He did his undergraduate and graduate work in physics at MIT. Subsequently he served on the staffs of Princeton University, Columbia, Cornell, and Stanford, before returning to MIT as a faculty member in 1981. Dr. Guth began his career working on abstract mathematical problems in elementary particle theory, but in 1978 he was persuaded by a friend to work on the application of particle theory to cosmology. This work led to the development of the "inflationary universe," a new model for the first fraction of a second of cosmic existence. This model and other aspects of cosmology are discussed in his chapter, "The Birth of the Cosmos."

Our second chapter concerns the formation of the Earth, one of the four terrestrial planets in our solar system. It is the only one of the four with intelligent life on it, and therefore the one that is particularly important to us. Astronomers know that there are 100 billion stars not too very different from the Sun in our galaxy, and that the number of galaxies within the reach of our telescopes has grown from about 100,000, estimated on the basis of an early photographic survey by James E. Keeler in 1900, to about one billion, a conservative present-day estimate. We have every reason to believe that even larger telescopes, looking out even further into the universe, would reveal even more galaxies. Hence, it is hard indeed to believe that there are *no* other solar systems somewhere out there, some of them with a planet not too different from the Earth, and perhaps people not too different from us—and perhaps even now one of them is reading a book on origins and extinctions!

Thus, the way in which the Earth formed may be a model of the way many other "earths" formed, in many other solar systems. Our chapter on this subject is by George W. Wetherill, director of the Department of Terrestrial Magnetism of the Carnegie Institution of Washington. Despite its name, DTM is in fact an outstanding geophysics and astronomy research center. Dr. Wetherill received his undergraduate and graduate training at the University of Chicago. He was deeply interested in astronomy from childhood. Beginning in experimental mass spectrometry and radioactive dating, Dr. Wetherill went on to theoretical work on the development of the orbits of objects in the solar system and on many other facets of planetary physics. In 1986 he was awarded the Gerard P. Kuiper

Prize of the Division of Planetary Sciences of the American Astronomical Society, in recognition of his outstanding contributions to the field.

We do not know when life originated on Earth, since there are few suitable rocks and no fossils from the first billion years of Earth history. For the past 3.5 billion years, however, we have a continuous record of life and atmospheric change. Anaerobic bacteria—those which do not directly use oxygen in their metabolism—dominated until about 2.5 billion years ago, when oxygenic, photosynthetic bacteria became more abundant and began to change the Earth's atmosphere rapidly to an oxygen-rich one. After more than a billion years (roughly 1.5 billion years ago), complex, eukaryotic organisms—protists, and ultimately animals and plants—evolved, and with them such modern attributes as multicellularity and sexuality.

Lynn Margulis, who has convinced most biologists of the symbiotic origin of eukaryotic cells by her extensive and well-reasoned analyses over the past two decades, is the author of our third chapter, "The Ancient Microcosm of Planet Earth." Like Dr. Wetherill, she did her undergraduate work at the University of Chicago, going on to graduate studies at the University of Wisconsin and the University of California, Berkeley. University Professor in the Department of Biology at Boston University, Dr. Margulis has made extensive contributions to our understanding of ancient life on Earth, in part because of her insightful studies of contemporary communities that share some of its characteristics.

Once it evolved, life on Earth has not simply developed in patterns of increasing complexity; rather, it has led a turbulent existence, challenged by environmental change. Multicellular organisms evolved in the sea somewhat more than 600 million years ago, adopting diverse life styles and forming interactive communities of often bewildering dimensions. The invasion of terrestrial habitats by plants, fungi, and animals began about 410 million years ago and led to an abundance of species in several major groups (insects and flowering plants, especially), although the greatest diversity of phyla still exists in the sea, where all but a very few apparently originated.

The evolution of life has been marked by spasmodic extinction events of enormous magnitude, the most recent and best-known of which occurred about 65 million years ago at the end of the Cretaceous period. That extinction, as well as several earlier ones, provided challenges to life on Earth that were generally met by new bursts of evolutionary diversification. For several years, scientists have been fascinated by the possibility that these extinctions were caused by cosmic events, such as the collision of major asteroids with the Earth, and that they may exhibit a certain periodicity.

No one has contributed more to our understanding of extinction than David M. Raup of the Department of the Geophysical Sciences at the University of Chicago. Dr. Raup explores this field in the fourth and final paper of this book, "Extinction in the Geologic Past." Like the other papers included here, this one attempts to establish patterns that may characterize other planets as well as our own. Dr. Raup received his undergraduate education at the University of Chicago and his doctorate at Harvard, serving on the faculties of the California Institute of Technology, Johns Hopkins University, and the University of Rochester before returning to Chicago in 1980.

This book is based on a symposium held at the National Academy of Sciences in Washington, D.C., on April 30, 1986. Its title was "Life and the Universe." These symposia are intended to cover scientific topics of wide interest to all the members of the academy, and this one was unusually successful in this respect. Bryce Crawford, Jr., the Home Secretary of the National Academy of Sciences from 1979 through 1987, originally conceived the idea of a symposium linking up the physical and biological sciences. He supported us in every stage of organizing it. To him this book is dedicated.

Donald E. Osterbrock
Peter H. Raven

Origins and Extinctions

ALAN H. GUTH

1   *The Birth of the Cosmos*

In about 1978 I joined a small drove of particle theorists who had begun to dabble in the early universe. We were motivated partly by the intrinsic fascination of cosmology, but also by developments in particle physics itself. The motivation arose primarily from the advent of a new class of particle theories known as grand unified theories. These theories were invented in 1974, but it was not until about 1978 that they became a topic of widespread interest in the particle physics community. The theories are spectacularly bold, attempting to extrapolate our understanding of particle physics to energies of about 10^{14} GeV (1 GeV = 1 billion electron volts $\approx$ rest energy of a proton). This amount of energy, by the standards of the local power company, may not seem so impressive—it is about what it takes to light a 100-watt bulb for a minute. The grand unified theories, however, attempt to describe what happens when that much energy is deposited on a single elementary particle. This extraordinary concentration of energy exceeds

Based on lectures given at the National Academy of Sciences (April 30, 1986), the Boston Museum of Science (April 1, 1987), and the Smithsonian Institution (April 8, 1987). This work was supported in part through funds provided by the United States Department of Energy under contract DE–AC02–76ERO3069, and in part by the National Aeronautics and Space Administration under grant NAGW–553.

1

the capabilities of the largest existing particle accelerators by 11 orders of magnitude.

To get some feeling for how high this energy really is, imagine trying to build an accelerator that might reach these energies. One can do it in principle by building a very long linear accelerator. The largest existing linear accelerator is the one at Stanford, which has a length of about 2 miles and a maximum energy of about 40 GeV. The output energy is proportional to the length, so a simple calculation shows how long the accelerator would have to be to reach an energy of 10^{14} GeV. The answer is almost exactly one light-year.

The United States Department of Energy, unfortunately, seems to be very unreceptive to proposals for funding a one-light-year accelerator. Consequently, if we want to see the most dramatic new implications of the grand unified theories, we are forced to turn to the only laboratory to which we have any access at all which has ever reached these energies. That "laboratory" appears to be the universe itself, in its very infancy. According to the standard hot big bang theory of cosmology, the universe had a temperature corresponding to a mean thermal energy of 10^{14} GeV at about 10^{-35} second after the big bang. So that is why particle theorists suddenly became interested in the very early universe.

The first half of this chapter will review the standard hot big bang model of the early universe; the second half will discuss the developments that have taken place since 1978, developments which have been motivated mainly by ideas from particle physics.*

THE BIG BANG THEORY

Cosmology in the twentieth century began with the work of Albert Einstein. In March 1916 Einstein completed a landmark paper titled "The Foundation of the General Theory of Relativity" (Einstein 1916). The theory of general relativity is in fact nothing more nor less than a new theory of gravity. It is a complex but very elegant theory, in which gravity is described as a distortion of the geometry of space and time. Unlike Newton's theory of gravity, general relativity is consistent with the ideas of "special relativity" that Einstein had introduced in 1905. While the rest of the world

---

*For the reader who would like a more detailed but still nontechnical treatment of these topics, I would recommend the discussion of the standard big bang model by Steven Weinberg (Weinberg 1977) and the discussion of more recent developments by John Gribbin (Gribbin 1986).

waited to be persuaded, Einstein was immediately convinced that he had found the correct description of gravity.

Immediately after his discovery of general relativity, Einstein proceeded to apply it to the universe as a whole. The results of his studies were submitted less than a year later, in February 1917 (Einstein 1917). In carrying out these studies, Einstein discovered something that surprised him a great deal: it is impossible to build a static model of the universe that is consistent with general relativity. Einstein was perplexed by this fact. Like his predecessors, he had looked into the sky, noticed that the stars appeared motionless, and erroneously concluded that the universe was static.

The problem that Einstein discovered in the context of general relativity also exists in Newtonian mechanics, although it was not appreciated until the work of Einstein. The problem is fairly simple to understand: if masses were distributed uniformly and statically throughout space, then everything would attract everything else and the entire configuration would collapse.

Einstein nonetheless remained convinced that the universe was obviously static. He therefore modified his equations of general relativity by adding what he called a "cosmological term," which amounts to a kind of universal repulsion that prevents the uniform distribution of matter from collapsing under the normal force of gravity. The cosmological term, Einstein found, fits neatly into the equations of general relativity; it is completely consistent with all the fundamental ideas on which the theory was constructed.

Einstein's ideas remained viable for about a decade, until astronomers began to measure the velocities of distant galaxies. They then discovered that the universe is not at all static. To the contrary, the distant galaxies are receding from us at high velocities.

The pattern of the cosmic motion was codified at the end of the 1920s by Edwin Hubble, in what we now know as Hubble's law. Hubble's law states that each distant galaxy is receding from us at a velocity that is, to a high degree of accuracy, proportional to its distance. Thus one can write

$$v = Hl,$$

where v is the recession velocity, l is the distance to the galaxy, and the quantity H is known as the Hubble "constant." I put the word *constant* in quotation marks to call attention to its inaccuracy. The quantity was called a "constant" by the astronomers, presumably because it remains approximately constant over the lifetime of an astronomer. The value of H changes as the universe evolves, however, so from the point of view of a cosmologist it is not a constant at all.

The value of the Hubble constant is not well known. The recession velocities of the distant galaxies are no problem—they can be determined very accurately from the Doppler shift of the spectral lines in the light coming from the galaxies. The distances to the galaxies, on the other hand, are very difficult to determine. These distances are estimated by a variety of indirect methods, and the resulting value of the Hubble constant is thought to be uncertain by a factor of about 2. It is believed to lie somewhere in the range

$$H \approx \frac{0.5 \text{ to } 1}{10^{10} \text{ years}}.$$

Notice that the Hubble constant has the units of inverse time; when the Hubble constant is multiplied by a distance, the result has the units of distance per time, or velocity. In particular, if the expression for H above is multiplied by a distance in light-years, the result is a velocity measured as a fraction of the velocity of light. Alternatively, one can use

$$H \approx (15 \text{ to } 30) \text{ km/sec per million light-years}$$

to obtain an answer in kilometers per second.

The development of cosmology in the twentieth century was somewhat confused by the fact that Hubble badly overestimated the value of the Hubble constant, reporting a value of 150 km/sec per million light-years. This mismeasurement had important consequences. In the context of the big bang model, an erroneously high value for the expansion rate implies an erroneously low value for the age of the universe. Hubble's value for the Hubble constant implied an age of about 2 billion years, a number which conflicts with geological evidence that the earth is significantly older. It was not until 1958 that the measured value of the Hubble constant came within the currently accepted range, due primarily to the work of Walter Baade and Allan Sandage.

Once it is noticed that the other galaxies are receding from us, there are two conceivable explanations. The first is that we might be in the center of the universe, with everything moving radially outward from us like the spokes on a wheel. In the early sixteenth century such an explanation would have been considered perfectly acceptable. Since the time of Copernicus, however, astronomers and physicists have become instinctively skeptical of this kind of

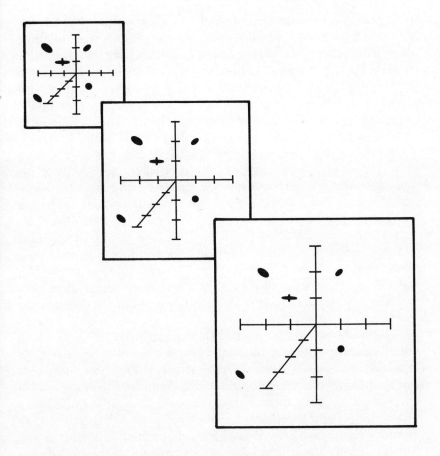

Fig. 1.1. *The expanding homogeneous universe.*

reasoning, so alternative explanations are sought. In this case an attractive alternative can be found.

The alternative explanation, illustrated in figure 1.1, can be called homogeneous expansion. The three diagrams are intended to show successive snapshots of a region of the universe. Each is essentially a photographic blowup of the previous diagram, with all distances enlarged by the same percentage. According to this explanation there is nothing special about our galaxy, or about any galaxy. All galaxies are approximately equivalent, so we can imagine that we might be living on any galaxy in the diagram. Galaxies are spread more or less uniformly throughout space; according to this description there is no center and no edge to the distribution of galaxies. As the system evolves from one diagram to the next, *all* intergalactic distances are enlarged. Thus, regardless of which galaxy we are living on, we will see all the other galaxies receding from us. Furthermore, this picture leads immediately to the conclusion that the recession velocities obey Hubble's law. Since all distances increase by the same *percentage* as the system evolves, the larger distances increase by a larger amount. The apparent velocity of a galaxy is proportional to the amount by which the distance from us increases, and hence it is proportional to the distance.

When one extrapolates this picture backwards in time, one finds that there is a certain instant in the past when the density of the universe would have been infinite. Such an event is called a singularity, and in this case the singularity is the instant of the big bang itself. The big bang occurred between 10 and 20 billion years ago. The time is very uncertain for two reasons: we do not know the Hubble constant very accurately and we are uncertain about the mass density of the universe. The mass density is important in calculating the history of the universe, because it determines how fast the cosmic expansion is slowing down under the influence of gravity.

The reader should be warned that the calculation implying an infinite density at the instant of the big bang is not to be trusted. As one looks backward in time with the density going up and up, one is led further and further from the conditions under which the laws of physics as we know them were developed. Thus, it is quite likely that at some point these laws become totally invalid, and then it is a matter of guesswork to discuss what happened at earlier times. However, one can still say that as the history of the universe is extrapolated backward using the laws of physics as we know them, the density increases without limit. Most cosmologists today are reasonably confident in our understanding of the history of the universe back to one microsecond (10^{-6} second) after the big bang, and the goal of the cosmological research involving grand unified theories is to solidify our understanding back to 10^{-35} second after the big bang.

Having discussed the key features of the big bang theory, we can now ask what evidence can be found to support it. In addition to Hubble's law, there are two significant pieces of observational evidence in favor of the theory. The first is the observation of the cosmic background radiation. To understand the origin of this radiation, begin by recalling that the temperature of a gas rises when the gas is compressed. For example, a bicycle tire is warmed when it is inflated by a hand pump. Similarly, a gas cools when it is allowed to expand. In the big bang model, the universe has been expanding throughout its history, and that means that the early universe must have been much hotter. (In fact, a mathematical treatment indicates that the temperature would have been infinite at the instant of the big bang. This infinity, however, like the infinite density, should not be considered convincing.)

Hot matter has the universal property that it emits a glow, just like the glow of hot coals in a fire. The hot matter of the early universe would therefore have emitted a glow of light that would have permeated the early universe. As the universe expanded this light would have red-shifted. Today the universe would still be bathed by the radiation, a remnant of the intense heat of the big bang, now red-shifted into the microwave part of the spectrum. This prediction was confirmed in 1964 when Arno A. Penzias and Robert W. Wilson of the Bell Telephone Laboratories (Penzias and Wilson 1965) discovered a background of microwave radiation with an effective temperature of about 3°K.

At the time of their discovery, Penzias and Wilson were not looking for cosmic background radiation. Instead they were searching for astronomical sources capable of producing low-level radio interference. They discovered a hiss in their detector which they carefully tracked down, verifying that it was an external source of radiation and not simply electrical noise from the receiver. They also found that the radiation arrived at the Earth uniformly from all directions in the sky. Later the spectrum of the radiation was measured and was found to agree exactly with the kind of thermal radiation that would be expected from the glow of hot matter in the early universe. A graph of the spectral content of the cosmic background radiation is shown in figure 1.2.

The second important piece of evidence supporting the big bang theory is related to calculations of what is called "big bang nucleosynthesis." This evidence is somewhat more difficult to understand than the cosmic background radiation, and it is therefore much less discussed in the popular scientific literature. To make any sense out of this argument, the reader must first understand that the big bang theory is not just a cartoon description of how the universe may have behaved; on the contrary, it is a very detailed model. Once one accepts the basic assumptions of the big bang theory, then knowledge of the laws of physics allows one to calculate how fast the universe would have expanded, how fast the expansion would have been slowed by gravity, how fast the universe would have cooled, and so on. Given this information, knowledge from nuclear physics

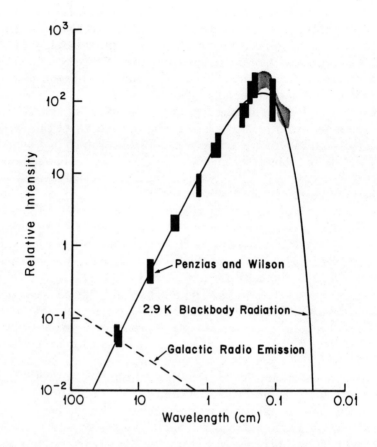

Fig. 1.2. *The spectrum of the cosmic background radiation. The vertical bars and the shaded region show the results of intensity measurements at a variety of wavelengths. The thin line is a theoretical curve, corresponding to thermal radiation at a temperature of 2.9°K.*

allows one to calculate the rates of the different nuclear reactions that took place in the early history of the universe.

The early universe was very hot, so hot that even nuclei would not have been stable. At two minutes after the big bang there were virtually no nuclei at all. The universe was filled with a hot gas of photons and neutrinos, with a much smaller density of protons, neutrons, and electrons. (The protons, neutrons, and electrons were very unimportant at the time, but later they became raw materials for the formation of stars and planets.) As the universe cooled, the protons and neutrons began to coalesce to form nuclei. From the nuclear reaction rates one can calculate the expected abundances of the different types of nuclei that would have formed. One finds that most of the matter in the universe would remain in the form of hydrogen. About 25% (by mass) of the matter would have been converted to helium, and trace amounts of other nuclei would also have been produced.

Most of the types of nuclei that we observe in the universe today were produced much later in the history of the universe, in the interiors of stars and in supernova explosions. The lightest nuclei, however, were produced primarily in the big bang, and it is possible to compare the calculated abundances with direct observations. Such a comparison can be carried out for the abundances of helium-4, helium-3, hydrogen-2 (otherwise known as deuterium), and lithium-7.

The comparison is complicated by the fact that we do not know all the information necessary to carry out the calculations. In particular, the calculation depends on the density of protons and neutrons in the universe, a quantity which can be estimated only roughly by astronomical observations. Thus, the calculations have to be carried out for a range of values for this density. One then asks whether there exists a plausible value for which the answers turn out right.

The results of this comparison are shown in figure 1.3. The horizontal axis shows the present density of protons and neutrons, and the curves indicate the results of the calculation. The observations, with the estimated range of their uncertainties, are shown by the shaded horizontal bars. Note that there is a range of values for the density of protons and neutrons, indicated by the cross-hatched region, for which each of the four calculated curves agrees with the corresponding observations. Even though the abundances of these nuclei are not known to high precision, the success of the comparison is very impressive. Notice that the abundances span 9 orders of magnitude. If there was never a big bang, there would be no reason whatsoever to expect that helium-4 would be 10^8 times as abundant as lithium-7—it might just as well have been the other way around. But this ratio can be calculated in the context of the big bang theory, and it works out just right.

UNANSWERED QUESTIONS

Each piece of evidence discussed above—Hubble's law, the cosmic background radiation, and the big bang nucleosynthesis calcula-

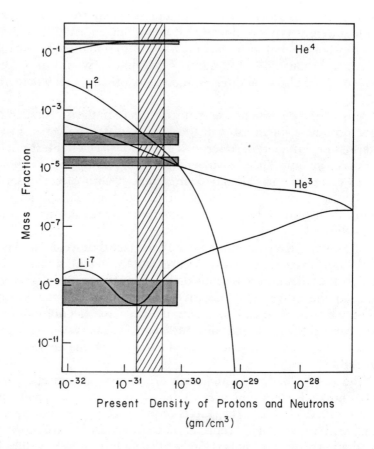

Fig. 1.3. *Big bang nucleosynthesis and the abundances of the light chemical elements. The curves show how the predicted abundances of helium-4, helium-3, deuterium (hydrogen-2), and lithium-7 depend on the present density of protons and neutrons. The shaded horizontal bars show the observed abundances, with the estimated uncertainties. Excellent agreement is obtained for the range of densities indicated by the cross-hatched region.*

tions—probes the history of the universe at a different period of time. The observation of Hubble's law, for example, probes the behavior of the universe at times comparable to the present, billions of years after the big bang. The cosmic background radiation, on the other hand, samples the conditions in the universe at about 100,000 years after the big bang, when the universe became cool enough for the plasma of free nuclei and electrons to condense into neutral atoms. The plasma that filled the universe at earlier times was almost completely opaque to photons, which would have been

TABLE 1.1   Questions Left Unanswered by the Standard Big Bang
Theory

1: Is there some way of understanding why the ratio of the number of photons to the number of protons and neutrons is about equal to 10^{10}, rather than some other number?

2: How did the universe become so homogeneous on large scales? Do we have to assume that it started out that way?

3: Why was the mass density of the early universe so extraordinarily close to the critical density?

4: Can one find a physical origin for the primordial density perturbations which lead to the evolution of galaxies and clusters of galaxies? Are there physical processes which determine the spectrum of these perturbations?

---

constantly absorbed and reemitted. With the formation of neutral atoms, however, the universe became highly transparent. Thus, most of the photons in the cosmic background radiation have been moving in a straight line since 100,000 years after the big bang, and they therefore provide an image of the universe at that time. Finally, the big bang nucleosynthesis calculations probe the history of the universe at much earlier times. The processes involved in determining the abundances of the light nuclei occurred at times ranging from about 1 second to about 4 minutes after the big bang.

The big bang theory is a very successful description of the evolution of the universe for the whole range of times discussed above, from about 1 second after the big bang to the present. Nonetheless, the standard big bang theory has serious shortcomings in that a number of very obvious questions are left unanswered. Here I will describe four of these questions, which are listed in table 1.1. Later I will show how new ideas from particle physics have led to a radically new picture for the very early behavior of the universe, a picture that provides plausible answers to each of these questions.

The first of the questions involves the number of protons and neutrons in the universe, relative to the number of photons. The photons are mainly in the cosmic background radiation, while the protons and neutrons form the atomic nuclei of the matter that makes up the galaxies. The observed universe contains about 10^{10} photons for every proton or neutron. The standard big bang theory does not explain this ratio, but instead assumes that the ratio is given as a property of the initial conditions.

The second question is related to the large-scale homogeneity, or uniformity, of the observed universe. The discussion of homogeneity must be qualified, however, because the universe that we observe is in many ways very inhomogeneous. The stars, galaxies, and clusters of galaxies make a very lumpy distribution. Cosmologically speaking, however, all of this structure in the universe is very small scale. If one averages over very large scales of 300 million light-years or more, then the universe appears to be very homogeneous. This large-scale homogeneity is most evident in the cosmic background radiation. Physicists have probed the temperature of the cosmic background radiation in different directions and have found it to be extremely uniform. It is just slightly hotter in one direction than in the opposite direction, by about 1 part in 1,000. Even this small discrepancy, however, can be accounted for by assuming that the solar system is moving through the cosmic background radiation at a speed of about 600 km/sec. Once the effect of this motion is subtracted out, the resulting temperature pattern is uniform in all directions to the best accuracy that has so far been attained—an accuracy better than 1 part in 10,000. Since the cosmic background radiation gives an image of the universe at 100,000 years after the big bang, one concludes that the universe was very homogeneous at that time. (The observed small-scale inhomogeneities are believed to have formed later, by the process of gravitational clumping.) The standard big bang theory cannot explain the large-scale uniformity; instead the uniformity must be postulated as part of the initial conditions.

The difficulty in explaining the large-scale uniformity is a quantitative question, related to the rate of expansion of the universe. Under many circumstances a uniform temperature would be easy to understand—anything will come to a uniform temperature if it is left undisturbed for a long enough period of time. In the standard big bang theory, however, the universe evolves so quickly that it is impossible for the uniformity to be created by any physical process. In fact, the impossibility of establishing a uniform temperature depends on none of the details of thermal transport physics, but instead is a direct consequence of the principle that no information can propagate faster than the speed of light. One can pretend, if one likes, that the universe is populated with little purple creatures, each equipped with a furnace and a refrigerator, and each dedicated to the cause of trying to create a uniform temperature. One can show by a straightforward calculation that the purple creatures would have to communicate at more than 90 times the speed of light in order to achieve their goal of creating a uniform temperature across the visible universe within 100,000 years after the big bang.

The puzzle of explaining why the universe appears to be uniform over such large distances is not a genuine inconsistency of the standard theory; if the uniformity is assumed in the initial conditions, then the universe will evolve uniformly. The problem is that one of the most salient features of the observed universe—its large-scale uniformity—cannot be explained by the standard big bang theory; it must be assumed as an initial condition.

The third question left unanswered by the theory is related to the mass density of the universe. This mass density is usually measured relative to a benchmark called the "critical mass density," which is defined in terms of the expansion of the universe. If the mass density exceeds the critical density, then the gravitational pull of everything on everything else will be strong enough to eventually halt the expansion. The universe will recollapse, resulting in what is sometimes called a "big crunch." If the mass density is less than the critical density, on the other hand, then the universe will go on expanding forever.

Cosmologists typically describe the mass density of the universe by a ratio designated by the Greek letter Ω (omega), defined as

$$\Omega \equiv \frac{\text{mass density}}{\text{critical mass density}}$$

Ω is very difficult to determine, but its present value is known to lie somewhere in the range of 0.1 to 2.

That seems like a broad range, but consideration of the time development of the universe leads to a different point of view. $\Omega = 1$ is an unstable equilibrium point of the evolution of the standard big bang theory, which means that it resembles the situation of a pencil balancing on end. The phrase *equilibrium point* implies that if Ω is ever exactly equal to 1, it will remain exactly equal to 1 forever—just as a pencil balanced precisely on end will in principle, according to the laws of classical physics, remain forever in the vertical position. The word *unstable* means that any deviation from the equilibrium point, in either direction, will rapidly grow. If the value of Ω in the early universe is just a little bit above 1, it will rapidly rise toward infinity; if Ω in the early universe is just a tiny bit below 1, it will rapidly fall toward zero. Thus, it seems very unlikely for the value of Ω today to lie anywhere in the vicinity of 1.

For Ω to be anywhere near 1 today, it must have been extraordinarily close to 1 at early times. For example, we can consider the time of 1 second after the big bang, the time at which the processes related to big bang nucleosynthesis were beginning to take place. In order for Ω to be somewhere in the allowed range today, at 1 second after the big bang Ω had to have been equal to 1 to an accuracy of 15 decimal places. If we go further and consider the time of 10^{-35} second after the big bang, the time when thermal energies were typical of the energy scale of grand unified theories, then at that time Ω had to have been equal to 1 to an accuracy of 49 decimal places!

In the standard big bang theory there is no explanation whatever for this fact, as has been emphasized by Robert H. Dicke and P. James E. Peebles of Princeton University (Dicke and Peebles 1979). At 1 second after the big bang Ω could have had any value, except that most possibilities would lead to a universe very different from the one in which we live. Like the large-scale homogeneity, the nearness of the mass density to the critical density cannot be explained, but must instead be postulated as part of the initial conditions.

The fourth question I wish to consider concerns the origin of the density perturbations that are responsible for the development of the small-scale inhomogeneities—the galaxies, clusters of galaxies, and so on. While the universe is remarkably homogeneous on the very large scales, there is nonetheless a very complicated structure on smaller scales. The existence of this structure is undoubtedly related to the gravitational instability of the universe: any region that contains a higher-than-average mass density will produce a stronger-than-average gravitational field, thereby pulling in even more excess mass. Thus, small perturbations are amplified to become large perturbations. Once these perturbations have begun, we understand more or less how they evolve, although the details get rather complicated.

However, in order for galaxies to evolve, the early universe must have contained primordial density perturbations. The standard big bang model offers no explanation for either the origin or the form of these perturbations. Instead an entire spectrum of primordial perturbations must be assumed as part of the initial conditions.

GRAND UNIFIED THEORIES

The four questions listed in table 1.1 involve some of the most basic and obvious features of the universe, yet the standard big bang theory leaves all of them unanswered. During the last decade, however, cosmologists have made use of new ideas from elementary particle theory to develop new ideas about the behavior of the universe at very early times. In the process they have discovered plausible answers to each of these questions. Before discussing these new ideas in cosmology, however, it is necessary to summarize the recent advances that have taken place in elementary particle physics.

Elementary particle physicists use the word *interaction* to refer to any process that elementary particles can undergo, whether it involves scattering, decay, particle annihilation, or particle creation. All of the known interactions of nature are divided into four types. From the weakest to the strongest, these interactions are gravitation, the weak interactions, electromagnetism, and the strong interactions. The force of gravity appears to be strong in our everyday lives because it is long-range and universally attractive. Thus, we are accustomed to feeling the force that acts between all the particles in the Earth and all the particles in our own bodies. The force of gravity acting between two elementary particles, however, is incredibly weak. It is much weaker than any of the other known forces, so weak, in fact, that it has never been detected. The weak interactions are much stronger than gravity, but they are not noticed in our everyday lives because they have a range that is roughly 100 times smaller than the size of an atomic nucleus. They are seen primarily in the radioactive decay of many kinds of nuclei and are also responsible for the scattering of particles called neutrinos, a type of experiment that is now routinely carried out at high-energy accelerator laboratories. Electromagnetism includes both electric and magnetic forces and is responsible for holding the electrons of an atom to the nucleus. Light waves, radio waves, microwaves, and X rays are also electromagnetic phenomena. The strong interactions, which have a range of about the size of an atomic nucleus, account for the force that binds the protons and neutrons inside a nucleus. They also account for the tremendous energy release of a hydrogen bomb, as well as the interactions of many short-lived particles that are investigated in particle accelerator experiments.

The strong, the weak, and the electromagnetic interactions all appear to be accurately described by theories developed during the early 1970s. The strong interactions are described by quantum chromodynamics, or QCD, a theory based on the hypothesis that all strongly interacting particles are composed of quarks. The theory provides a detailed description of the interactions that bind the quarks into the observed particles, and the residual effect of these quark interactions can account for the observed interactions of the particles. Unfortunately, our ability to extract quantitative predictions from QCD is very limited. The theory is very intricate and at present only some of its consequences can be reliably calculated. Nonetheless, the evidence for QCD is strong enough that most particle physicists are convinced that the theory correctly describes the strong interactions over the full range of available energies.

The weak and electromagnetic interactions are successfully described by the unified electroweak theory, also known as the Glashow-Weinberg-Salam model (named for Sheldon Lee Glashow of Harvard University, Steven Weinberg of the University of Texas at Austin, and Abdus Salam of the International Center for Theoretical Physics in Trieste, who shared the 1979 Nobel Prize in physics for this work). Standard calculational techniques are very effective in extracting predictions from this theory, owing to the inherent weakness of the interactions being described.

While a quantum theory of gravity remains to be developed, we nonetheless believe that general relativity provides the correct description of gravity at the level of classical physics—that is, in the approximation that the effects of quantum theory can be ignored. The effects of gravity, however, are noticeable only when the number of elementary particles is very large. The classical approximation is incredibly accurate in these situations, and the theory of general relativity is therefore sufficient to describe all the observed properties of gravity.

Quantum chromodynamics and the unified electroweak theory, when taken together, have come to be called the standard model of elementary particle physics. Embedded in the standard model are three different types of fundamental interactions, labeled by the symbols $U(1)$, $SU(2)$, and $SU(3)$. (These symbols are actually the names of mathematical symmetry groups which determine the form of the interactions, but for our purposes the symbols can be taken simply as labels for the three interactions.) The $U(1)$ and $SU(2)$ interactions are the fundamental ingredients of the Glashow-Weinberg-Salam theory, and they combine together in a somewhat complicated way to describe the weak and electromagnetic interactions. The $SU(3)$ label refers to the strong interactions described by quantum chromodynamics.

Thus, since the early 1970s elementary particle physics has been in a state of unprecedented success. The electromagnetic, weak, and strong interactions are successfully described by the standard model of particle physics in terms of three fundamental interactions. It appears that all known physics can be described by the standard model of particle physics and/or the theory of general relativity.

Grand unified theories—often referred to by their acronym, GUTs—emerged from this atmosphere of enormous success. The first grand unified theory, called the "minimal $SU(5)$ model," was proposed in 1974 by Glashow and Howard Georgi, also of Harvard University (Georgi and Glashow 1974).

The basic idea of grand unification is that the $U(1)$, $SU(2)$, and $SU(3)$ interactions of the standard model of particle physics are actually components of a single unified force. At first this idea seems impossible, since the strengths of the three types of interactions are very different. The interactions strengths cannot be determined theoretically, but instead must be fixed by experiment. The theory, however, implies that the interaction strengths depend on the energy of the particles that are interacting. Once the strengths of the three interactions are measured at one energy, the theory allows one to calculate the strengths at any other energy. The results of such a calculation are shown in figure 1.4. The important feature is that all three lines appear to meet rather accurately at a single point at an energy somewhere between 10^{14} and 10^{15} GeV. It is this calculation, first carried out by Georgi, Weinberg (then at Harvard University), and Helen R. Quinn (then at Harvard University, now at the Stanford Linear Accelerator Center), that determines the enormous energy scale of the grand unified theories (Georgi, Quinn, and Weinberg 1974).

According to the grand unified theories, there is really only one interaction, not three. If we were able to do experiments in the energy range of 10^{14} or 10^{15} GeV, then we would see clearly that there is only one interaction. At lower energies, however, the theory contains a mechanism that causes the one interaction to look as if it were three interactions. The mechanism is called spontaneous symmetry breaking. I will not be able in this chapter to explain fully how spontaneous symmetry breaking works, but I will return in the next section to describe in more detail some of its properties. For now, the reader should be aware that the mechanism of spontaneous symmetry breaking is not new with grand unified theories. The mechanism has been used very successfully in the Glashow-Weinberg-Salam theory of the electroweak interactions, and similar phenomena are known to occur in condensed matter physics.

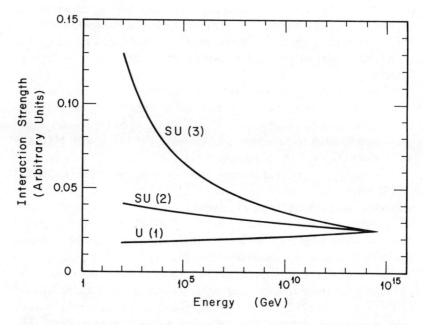

Fig. 1.4. *Dependence of interaction strengths on energy. The strengths of the three types of fundamental interactions—U(1), SU(2), and SU(3)—are measured at energies of about 100 GeV. The curves show the calculated strengths of these interactions at higher energies according to the standard model of particle physics. The meeting of all three curves at an energy between 10^{14} and 10^{15} GeV suggests that the three interactions are unified at this energy.*

While the standard model of particle physics discussed earlier is very well established, the same cannot be said for grand unified theories. Even if the idea of grand unification is correct, we certainly do not know which of the many conceivable grand unified theories is likely to be the right one. Nonetheless, grand unified theories are considered highly attractive for a variety of reasons, and here I would like to explain two of them.

First of all, the grand unified theories are the only known theories which predict that the charges of the electron and the proton should be equal in magnitude. When I tell this to my friends who are not particle physicists, many of them are unimpressed. They learned in high school that those two charges are equal and it never seemed very consequential. High school teachers are notorious, however, for neglecting to tell their students that prior to grand unified theories, nobody had even a fuzzy idea about *why* those two charges are equal. In all theories that had been developed prior to the grand unified theories, the two charges could each have been any number whatever; it was just an experimental coincidence that for some reason they happened to be equal to each other to at least 1 part in 10^{20}. Grand unified theories, on the other hand, contain a fundamental symmetry that relates the behavior of electrons to the behavior of the quarks that make up the proton. This symmetry guarantees that the charges are equal. Furthermore, if this symmetry were violated by even the smallest amount, then the theory would no longer be mathematically well defined. Thus, if the charge of the electron were found to differ in magnitude from the charge of a proton by, for example, 1 part in 10^{24}, then the grand unified theories would have to be abandoned. But if successively more accurate experiments continue to confirm that the two charges are equal, such a result would have to be considered as further evidence in favor of grand unification.

A second reason why grand unified theories are considered attractive involves a topic that I have already discussed, but that I will now describe more quantitatively. Grand unified theories, based on the idea that the three interactions of the standard model of particle physics arise from a single fundamental interaction, imply that the three curves describing the interaction strengths in figure 1.4 must all meet at a point. That means, for example, that if any two of the interaction strengths are measured, then the third interaction strength can be predicted by the criterion that its curve must pass through the point where the first two curves crossed each other. This prediction of the grand unified theories is found to work very well—the experimental result agrees with the prediction to an accuracy of about 3%, while the estimated experimental uncertainty of the test is about 6% (see, for example, Gasiorowicz and Rosner 1981).

To present the full picture, I should mention that grand unified theories suffer from one important drawback, known as the "hierarchy problem." This problem is largely aesthetic, but it is taken quite seriously by the particle physics community. The problem is that the enormous energy scale of the grand unified theories—10^{14} GeV—has to be "put in by hand." When we say the number is "put in by hand," we mean that there is no known a priori reason why this energy scale is so many orders of magnitude larger than the other energy scales of importance to particle physics. Recall, however, that there is a clear *experimental* reason for believing that the energy scale of unification is very high: according to figure 1.4, the scale of unification must be very high in order to account for the large differences in the strengths of the three interactions observed at the energies of experimental particle physics. To understand the attitude of the particle theorists, one must realize that the grand unified theories are not seen as the ultimate fundamental theory of nature. First of all, the ultimate fundamental theory must obviously include gravity, which grand unified theories do not. Second, the grand unified theories are considered too inelegant to be serious candidates for the ultimate fundamental theory of nature. In particular, even the simplest of the grand unified theories contains over 20 free parameters (that is, numbers, such as the charge of an electron, that must be measured experimentally before the theory can be used to make predictions). Thus, the particle theorist expects that some day the correct grand unified theory will be derived as an approximation to the ultimate theory, which will contain few if any free parameters. In this context, the energy scale of grand unification will be calculable. When the particle theorist says that the energy scale of grand unified theories is "put in by hand," he is really saying that he does not at present see any reason why this hypothetical calculation of the future will give such a large number. So the advocates of grand unified theories are hoping that someday the reason will be found.

SPONTANEOUS SYMMETRY BREAKING

Before returning to the discussion of cosmology, I would first like to describe some of the properties of spontaneous symmetry breaking. According to the general definition, a spontaneously broken symmetry is one which is present in the underlying theory describing a system, but which is hidden when the system is in its equilibrium state. In order to give the reader some familiarity with this concept, I will try to explain the spontaneous symmetry breaking in a grand unified theory by comparing it with the spontaneous symmetry breaking that occurs in a much more familiar system—a crystal. This section will nonetheless be somewhat more complicated than the rest of the chapter, so the reader who finds it tedious is invited to skip to the next section. The discussion of spontaneous symmetry breaking will not be necessary to understand what will follow, although the reader who is willing to wade through this section will certainly finish up with a more complete understanding.

The analogy between spontaneous symmetry breaking in crystals and spontaneous symmetry breaking in grand unified theories

TABLE 1.2 Spontaneous Symmetry Breaking: The Crystal-GUT Analogy

	Crystal	GUTS
Symmetry	Rotational invariance.	Electron, neutrino, and quark indistinguishable. Three interactions indistinguishable.
Spontaneous Symmetry Breaking	Crystal axes pick out three distinguished directions.	Higgs fields pick out three distinguished particles—electron, neutrino, and quark—and also three distinguished interactions.
Low-Energy Physics	Three fundamental axes of space. Three fundamental speeds of light.	Three distinguished particles. Three distinguished interactions.
High-Temperature Physics	Crystal melts. Rotational invariance restored.	Phase transition at $T \approx 10^{27}$ °K. Symmetry restored.

is outlined in table 1.2. In order to make the analogy as clear as possible, I will discuss a particularly simple type of crystal, a type called "orthorhombic." The structure of an orthorhombic crystal is illustrated in figure 1.5. These crystals have a rectangular structure, so all the angles are right angles. However, unlike a simple cubic crystal in which all the principal lengths are equal, the three principal lengths of orthorhombic crystals are all different. (This feature of orthorhombic crystals will make the analogy with grand unified theories a little closer.) A crystal of topaz provides an example of the orthorhombic structure.

Starting at the top of table 1.2, the first row indicates the symmetry that is involved. In the case of the crystal, the relevant symmetry is rotational invariance—that is, the physical laws that describe the system are rotationally invariant, in that they make no distinction between one direction of space and another. In the case of the grand unified theory, the symmetry is more abstract, having

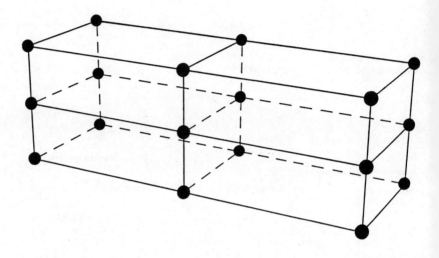

Fig. 1.5. *The structure of an orthorhombic crystal. This type of crystal has a rectangular structure in which all three principal lengths are different. The formation of such a crystal is an example of spontaneous symmetry breaking, analogous to the spontaneous symmetry breaking in grand unified theories.*

nothing to do with orientation in physical space. Instead the symmetry is what is sometimes called an "internal" symmetry, one that relates the behavior of one type of particle to the behavior of another. In this case, the underlying symmetry of the grand unified theory implies that the three interactions of the standard model of particle physics—$U(1)$, $SU(2)$, and $SU(3)$—are really one interaction and hence indistinguishable. In addition, the symmetry implies that particles which are normally distinguished from each other by the way in which they participate in these interactions will necessarily lose their identities. In particular, the grand unified theories imply that the underlying laws of physics make no distinction between an electron, a neutrino, and a quark. The indistinguishability of these three particles, as well as the indistinguishability of the $U(1)$, $SU(2)$, and $SU(3)$ interactions, is analogous to the indistinguishability of the different directions of space in the case of the crystal.

The second row of table 1.2 indicates the nature of the spontaneous symmetry breaking. In the case of the crystal, the atoms arrange themselves along specific crystallographic axes that are picked out randomly as the crystal is formed, and thus three directions of space become distinguished. In grand unified theories, a set of fields is added for the specific purpose of spontaneously breaking the symmetry. These fields are known as Higgs fields (after Peter W. Higgs of the University of Edinburgh) and the spontaneous symmetry-breaking mechanism, which occurs in a variety of particle physics theories, is known as the Higgs mechanism.

In terms of the structure of the theory, the Higgs fields are on an equal footing with the other fundamental fields, such as the electromagnetic field. It is postulated that these fields exist and that they evolve according to a specified set of equations. While the electromagnetic field gives rise to photons, the Higgs fields give rise to Higgs particles. The Higgs particles associated with the breaking of the grand unified symmetry are expected to have masses corresponding to energies in the vicinity of 10^{14} GeV, which means that they are far too massive to be produced in the foreseeable future. There is another Higgs particle, however, associated with the spontaneous symmetry breaking in the Glashow-Weinberg-Salam theory, which could conceivably be detected in upcoming accelerator experiments.

The spontaneous symmetry breaking is accomplished by formulating the theory in such a way that the Higgs fields have nonzero values in the vacuum. (To the particle theorist, the word *vacuum* is defined to be the state of lowest possible energy density.) Just as the atoms in the crystal can align equally well along any of an infinite number of possible orientations, the Higgs fields can assume any of an infinite number of combinations of values. Some particular combination of Higgs field values is chosen randomly as the system is formed, just as distinguished directions in space are chosen randomly as a crystal begins to form. This random choice of nonzero Higgs field values breaks the grand unified symmetry. The other particles in the theory interact with the Higgs fields, producing the apparent distinction between the $U(1)$, $SU(2)$, and $SU(3)$ interactions, and also the apparent distinction between electrons, neutrinos, and quarks.

The third row of table 1.2 describes the behavior of low-energy physics in the two systems. Here the analogy can be made more illustrative by whimsically imagining a world of intelligent creatures living inside an orthorhombic crystal. Let us assume that these creatures can somehow move about and carry on the task of scientific investigation, but that they cannot muster enough energy to melt or even significantly perturb the crystal in which they live.

In this world the crystal would not be considered an object, but instead the crystalline structure would be taken as a fundamental property of space. A physics book would make no mention of rotational symmetry, but would instead contain a chapter discussing the properties of space and its primary axes. The crystalline structure would, for example, affect the propagation of light through the medium, and a table of physical constants in the crystal world would list three speeds of light, one for each primary axis. If grand unification is correct, then the world of our experience is similar to this crystal world; our tabulation of the different properties of the strong, weak, and electromagnetic interactions is analogous to the tabulation of the different speeds of light. Similarly, the distinct properties that we observe for electrons, neutrinos, and quarks are not fundamental—they represent the different ways that particles can interact with the fixed "Higgs field crystal" in which we live.

The last row of table 1.2 describes the high-temperature behavior of the two systems. If a crystal is heated sufficiently, it will undergo a phase transition (that is, it will melt) to become a liquid. The distribution of molecules in the liquid is rotationally symmetric, looking the same no matter how the liquid is turned. Thus, at high temperatures, the rotational symmetry is restored. According to grand unified theories, an analogous phase transition occurs at a temperature of the order of $10^{27}°K$. (This temperature corresponds to a mean thermal energy of—you guessed it—10^{14} GeV.) At temperatures higher than this value the "Higgs field crystal" in which we live would enter a different phase. The Higgs fields would oscillate wildly under the thermal agitation, but the mean value of each would be zero and the grand unified symmetry would be restored. In this phase the $U(1)$, $SU(2)$, and $SU(3)$ interactions would all merge into a single interaction and there would be no distinction whatever between electrons, neutrinos, and quarks. If we lived at $10^{27}°K$, the concept of grand unification would be commonplace, rather than novel.

A temperature of $10^{27}°K$ is of course outrageously large, even by the standards of astrophysics. The center of a hot star, for example, is only about $10^7°K$. The application of grand unified theories, however, forces us to consider such outrageous temperatures. Since the cosmological consequences of the grand unified theories seem very attractive, one gains some confidence that our understanding of physics at these temperatures has a reasonable chance of being at least on the right track.

THE BARYON NUMBER OF THE UNIVERSE

Having discussed the particle physics background, I will now re-turn to the four questions that were introduced in table 1.1. While these questions were left totally unanswered by the standard big bang model, I will now be able to show how the new ideas from particle physics can provide plausible answers to each of them.

I will begin with the first of the questions—why the ratio of the number of photons to the number of protons and neutrons is about equal to 10^{10}. The idea that particle physics could provide an answer to this question was suggested first by Andrei D. Sakharov (Sakharov 1967), and the more detailed calculation in the context of grand unified theories was first carried out by Motohiko Yoshimura of Tohoku University (Yoshimura 1978) and by Weinberg (Weinberg 1979). The study of this question was the first application of grand unified theories to cosmology, and the subject remains crucial to our understanding of cosmology in this context.

Particle physicists use the word *baryon* to refer to either a proton or a neutron. More precisely, particle physicists define the "baryon number" of a system by the equation

baryon number ≡ (number of protons) + (number of neutrons)
  − (number of antiprotons) − (number of antineutrons) + . . . ,

where " . . . " denotes the contributions from other particles which are very short-lived, and therefore irrelevant to the questions we are now discussing.

It is useful to have a single word to refer to either a proton or a neutron, because in the early universe the two types of particles rapidly interconverted, by processes such as

proton + electron ↔ neutron + neutrino.

The baryon number is left unchanged by the reaction above, since the proton and the neutron each have a baryon number of 1. In fact, all physical processes observed up to now obey the principle of baryon number conservation: the total baryon number of an isolated system cannot be changed. This principle implies, for example, that the proton must be absolutely stable; because it is the lightest baryon, it cannot decay into another particle without changing the total baryon number. Experimentally, the lifetime of the proton is now known to exceed 10^{32} years.

Note, however, that the principle of baryon number conservation does not forbid the production of baryons, provided that equal numbers of antibaryons are also produced. For example, at high energies the reaction

$$\text{electron} + \text{positron} \rightarrow \text{proton} + \text{antiproton}$$

is frequently observed.

To estimate the baryon number of the observed universe, one must ask whether the distant galaxies are composed of matter or whether some might perhaps be formed from antimatter. This question has not been definitively answered, but there is a strong consensus that the universe is probably made entirely from matter. The belief is motivated mainly by the absence of any known mechanism that could have separated the matter from the antimatter over such large distances. Assuming that this belief is true, then the total baryon number of the visible universe is about 10^{78}, corresponding to about 1 baryon per 10^{10} photons.

If the principle of baryon number conservation were absolutely valid, then the baryon number of the universe would be unchangeable. Under this assumption there would be no hope of explaining the baryon number of the universe—it would always have had a value of 10^{78}, a value that was necessarily fixed by the postulated initial conditions of the universe.

Grand unified theories, however, imply that baryon number is *not* exactly conserved. At low temperatures the conservation law is an excellent approximation, and the observed limit on the proton lifetime is consistent with at least many versions of grand unified theories. At temperatures of $10^{27}°K$ and higher, however, processes that change the baryon number of a system of particles are expected to be quite common.

Thus, when grand unified theories and the big bang picture are combined, the net baryon number of the universe can be altered by baryon nonconserving processes. However, in order to explain the observed baryon number, it is necessary that the underlying particle physics make a distinction between matter and antimatter. This distinction is essential, since any theory that leads to the production of matter and antimatter with equal probabilities would lead to a total baryon number far smaller than what is observed. For many years it was thought that matter and antimatter behave identically, but a small difference between the two was discovered experimentally in 1964 by Val L. Fitch of Princeton University and James W. Cronin of the University of Chicago (who shared the 1980 Nobel Prize in physics for this discovery). This inherent distinction between matter and antimatter has since been incorporated into many particle theories, including the grand unified theories. Thus, in the context of grand unified theories, the observed excess of matter over antimatter can be produced naturally by elementary-particle interactions at temperatures near $10^{27}°K$.

Finally, then, we come to the crucial question: do grand unified theories give an accurate prediction for the baryon number of the observed universe? Unfortunately, we cannot tell. The grand unified theories depend on too many unknown parameters to allow a quantitative prediction. However, one can say that the observed baryon number can be obtained with what seems to be a reasonable choice of values for these unknown parameters. Thus, grand unified theories provide at least a framework for answering the first question of table 1.1. Sometime in the future, if the correct grand unified theory and the values of its free parameters become known, it will be possible to make a real comparison between theory and observation.

THE INFLATIONARY UNIVERSE

The answers that I will discuss for the three remaining questions from table 1.1 all depend on a new model for the very early behavior of the universe, a model called the inflationary universe. Before discussing the answers, I will have to spend some time describing how the model works. The description will be somewhat sketchy, but I will try to explain the main features.*

The inflationary universe was first proposed in 1981 (Guth 1981), but the model in its original form did not quite work. It had a crucially important technical flaw, which was pointed out but not remedied in the original paper. A variation that avoids this flaw was invented independently by Andrei D. Linde of the P. N. Lebedev Physical Institute in Moscow (Linde 1982) and by Andreas Albrecht and Paul J. Steinhardt of the University of Pennsylvania (Albrecht and Steinhardt 1982). This chapter will discuss the Linde-Albrecht-Steinhardt version of the model, which is called the new inflationary universe.

The key ingredient of the inflationary universe model is the

*For the reader who would like a more detailed but still nontechnical treatment, I would recommend the article by Edward P. Tryon (Tryon 1987) or the article by Paul J. Steinhardt and me (Guth and Steinhardt 1987). There are also a variety of technical reviews in the literature: Blau and Guth 1987; Brandenberger 1985; Linde 1984; Steinhardt 1986; and Turner 1987. Finally, there is at least one volume of collected papers: Abbott and Pi 1986.

assumed occurrence of a phase transition in the very early history of the universe. Grand unified theories imply that such a phase transition occurred when the temperature was about $10^{27\circ}$K (see table 1.2). This phase transition is linked to the spontaneous symmetry breaking: at temperatures higher than $10^{27\circ}$K there is one unified type of interaction, while at temperatures below $10^{27\circ}$K the grand unified symmetry is broken, and the $U(1)$, $SU(2)$, and $SU(3)$ interactions acquire their separate identities.

When the universe cooled down to the temperature of this phase transition, either of two things may have happened: the phase transition may have occurred immediately or it may have been delayed, occurring only after a large amount of supercooling. The word *supercooling* refers to a situation in which a substance is cooled below the normal temperature of a phase transition, without the phase transition taking place. Water, for example, can be supercooled to more than 20°K below its freezing point, and glasses are formed by rapidly supercooling a liquid to a temperature well below its freezing point. If the correct grand unified theory and the values of its parameters were known, there would be no ambiguity about the nature of the phase transition; we would be able to calculate how quickly it would occur. In the absence of this knowledge, however, either of the two possibilities appears plausible. Calculations show, however, that only an extremely narrow range of parameters leads to an intermediate situation; in almost all cases the phase transition is either immediate or strongly delayed.

If the phase transition occurred immediately, then its cosmological consequences would be very problematical. In that case a large number of exotic particles called magnetic monopoles would be produced, and the mass density of the universe would come to be strongly dominated by these particles. According to most grand unified theories, these monopoles would survive to the present day, leading to predictions that are grossly at odds with observation. Furthermore, in the case of an immediate phase transition the last three questions of table 1.1 would all remain unanswered.

The inflationary universe model is based on the other possibility, that the universe underwent extreme supercooling. The cosmological consequences of this assumption appear to be very attractive.

As the gas that filled the universe supercooled to temperatures far below the temperature of the phase transition, it would have approached a very peculiar state of matter known as a false vacuum. This state of matter has never been observed. Furthermore, the energy density required to produce it is so enormous—about 60 orders of magnitude larger than the density of the atomic nucleus—that it clearly will not be observed in the foreseeable future. Nonetheless, from a theoretical point of view the false vacuum seems to be well understood. The essential properties of the false vacuum depend only on the general features of the underlying particle theory and not on any of the details. Even if grand unified theories turn out to be incorrect, it is still quite likely that our theoretical understanding of the false vacuum would remain valid.

The false vacuum has a peculiar property that makes it very different from any ordinary material. For ordinary materials, whether they are gases, liquids, solids, or plasmas, the energy density is dominated by the rest energy of the particles of which the material is composed. (The rest energy of these particles is related to their masses by the famous Einstein relation, $E = mc^2$.) If the volume of an ordinary material is increased, then the density of particles decreases, and therefore the energy density also decreases. The false vacuum, on the other hand, is the state of lowest possible energy density that can be attained while remaining in the phase for which the grand unified symmetry is unbroken. This energy density is attributed not to particles, but rather to the Higgs fields which are responsible for the spontaneous symmetry breaking. Recall that we are assuming that the phase transition occurs very slowly, so for a long time (by the standards of the very early universe) the false vacuum is the state with the least possible energy density that can be attained. Thus, even as the universe expands, the energy density of the false vacuum remains constant.

When this peculiar property of the false vacuum is combined with Einstein's equations of general relativity, one finds a very dramatic result: the false vacuum leads to a gravitational repulsion. Throughout the rest of the history of the universe, gravity has acted to slow down the cosmic expansion, but when the universe was caught in the false vacuum state, gravity actually caused the expansion to accelerate. The form of this repulsion is identical to the effect of Einstein's cosmological constant, except that the repulsion caused by the false vacuum operates for only a limited period of time. Unfortunately, I have not found a way to convincingly explain the gravitational repulsion without invoking the detailed mathematics of general relativity, but in the next section I will describe in general terms how the repulsion arises.

The gravitational repulsion would have produced a very rapid expansion, far in excess of the expansion of the standard big bang model. In the inflationary model essentially all the momentum of the big bang was produced by the gravitational repulsion. (In the standard big bang theory, by contrast, all the momentum of the big bang is incorporated into the postulated initial conditions.) The universe would double in size in about 10^{-34} second, and it would continue to double in size during each successive interval of 10^{-34} second for as long as the universe remained in the false vacuum state. During this period the universe expanded, or "inflated," by a stupendous factor. A factor of at least 10^{75} (in volume) is necessary to answer the cosmological questions in table 1.1, but the actual number depends on the highly uncertain details of the underlying particle theory and may have been many orders of magnitude larger.

Eventually the phase transition would have occurred, and when it did the energy density of the false vacuum would have been released. (In the language of thermodynamics, this energy is the "latent heat" of the phase transition.) This energy input would have produced a vast number of particles and would have reheated the universe back to a temperature that is comparable to the temperature of the phase transition: about $10^{27}°K$. (The precise number is a factor of 2 or 3 below the temperature of the phase transition, but such factors go unnoticed in the order of magnitude estimates.) The baryon-number-producing processes discussed in the previous section would have taken place during and just after the reheating— any baryon number density present before inflation would have been diluted to a negligible value by the enormous expansion. At the end of the phase transition the universe would have been uniformly filled with a hot gas of particles, exactly as had been postulated as the initial condition for the standard big bang theory. Here the inflationary model merges with the standard big bang theory; the two models agree in their description of the evolution of the universe from this time onward.

In the inflationary model, virtually all the matter and energy in the universe were produced during the inflationary process. This seems strange, because it sounds like an unmistakable violation of the principle of energy conservation. How could it be possible that all the energy in the universe was produced as the system evolved?

The inflationary universe model is consistent with all the known laws of physics, including the conservation of energy. The loophole in the conservation of energy argument is associated with the peculiar nature of gravitational energy. Using either general relativity or Newtonian gravity, one finds that *negative* energy is stored in the gravitational field. A simple way to make this fact seem at least plausible is to imagine two large masses, separated by a very large distance in an otherwise empty space. Now imagine bringing the two masses together. The masses will attract each other gravitationally, which means that energy can be extracted as the masses come together. One can imagine, for example, attaching the masses to fixed pulleys, and the wheels of the pulleys can be used to drive an electrical generator. Once the two masses are brought together, however, their gravitational fields will be superimposed, producing a much stronger gravitational field. Thus, the net result of this process is to extract energy and produce a stronger gravitational field. If energy is conserved, then the energy in the gravitational field apparently goes down when its strength goes up. If the absence of a gravitational field corresponds to no energy, then any nonzero field strength must correspond to a negative energy. Gravitational energy is usually negligible under laboratory conditions, but cosmologically it can be very significant.

The energy stored in the false vacuum became larger and larger as the universe inflated, and was then released when the phase transition took place at the end of the inflationary period. At the same time, however, the energy stored in the cosmic gravitational field—the field by which everything in the universe is attracting everything else—became more and more negative. The total energy of the system was conserved, remaining constant at a value at or near zero.

Thus, inflation allows the entire observed universe to develop from almost nothing. The inflationary process could have started with an amount of energy equivalent to only about 10 kilograms of matter, and even this small amount of energy could conceivably have been balanced by an equal contribution of negative energy in the gravitational field. Thus, if the inflationary model is correct, it is fair to say that the universe is the ultimate free lunch.

NEGATIVE PRESSURE AND GRAVITATIONAL REPULSION

The mechanism that drives the accelerated expansion of the inflationary model cannot be described in detail without the formalism of general relativity, but in this section I will try to explain crudely how the gravitational repulsion arises. The material in this section is unnecessary for the rest of the chapter, so some readers may wish to skip immediately to the next section.

Recall that the false vacuum is the state of lowest possible energy density that can be attained while remaining in the phase for which the grand unified symmetry is unbroken. This energy density is attributed to the Higgs fields. In the vacuum state the Higgs fields have nonzero values that break the grand unified symmetry, and in this state the energy density of the Higgs fields is zero, or at least very small. In the false vacuum, on the other hand, each Higgs field has a value of zero, preserving the grand unified symmetry. In order to achieve the spontaneous symmetry breaking, however, the theory was formulated so that the state in which all the Higgs fields vanish is a state of high energy density! (Although it seems strange that energy should be required in order for the value of the Higgs fields to be zero, particle physicists find that this property causes no inconsistencies, and is exactly what is needed to produce the spontaneous symmetry breaking.) Thus, as a region of false vacuum expands, the energy density remains constant—it is just the energy density necessary to maintain a value of zero for the Higgs fields.

The constancy of the energy density of the false vacuum is related to another very peculiar property: the false vacuum has a pressure that is large and negative. To understand the connection between these two properties, consider the fact that when a normal, positive pressure gas is allowed to expand, it will push on its surroundings and in the process will lose energy to its surroundings. Both steam and gasoline engines operate on this principle. For the false vacuum, however, the situation is reversed. We can imagine a region of false vacuum that expands, but the expansion occurs at a constant energy density. The energy of the false vacuum region therefore increases as the volume increases, which means that the region is taking energy from its surroundings. This indicates that the region must create a negative pressure, or suction, so that energy is being supplied by whatever force is causing the expansion. By considering the energy balance involved in the expansion of a region of false vacuum, it is possible to determine the pressure uniquely. The pressure is equal to the negative of the energy density, when the two are measured in the same units.

According to Newton's theory of gravity, a gravitational field is produced by a mass density. In a relativistic theory the mass density can be related to a corresponding energy density by $E = mc^2$. According to Einstein's theory of general relativity, however, a pressure can also produce a gravitational field. When Einstein's equations are used to describe a homogeneously expanding universe, they show that the rate at which the expansion is slowed down is proportional to the energy density plus 3 times the pressure. Under ordinary circumstances the pressure term is a small relativistic correction, but for the false vacuum the pressure term overwhelms the energy-density term and has the opposite sign. So the bizarre notion of negative pressure leads to the even more bizarre effect of a gravitational force that is effectively repulsive.

ANSWERS TO THE REMAINING QUESTIONS

Having described the foundations of the inflationary universe model, I can now explain how the remaining questions of table 1.1 can be resolved. First I will discuss the second question, concerning the large-scale homogeneity of the universe. Recall that in the standard big bang theory, the large-scale homogeneity cannot be explained because the universe did not have enough time to come to a uniform temperature.

Consider now the evolution of the observed region of the universe, which has a radius today of about 10 billion light-years. Imagine following this region backward in time, using the inflationary model. Follow it back to the instant immediately before the inflationary period. Since the theory predicts a tremendous spurt of expansion during the inflationary period, one infers that the region was incredibly small before this expansion began. In fact the region was more than a billion times smaller than the size of a proton. (Note that I am *not* saying that that universe as a whole was very small. The inflationary model makes no statement about the size of the universe as a whole, which might in fact be infinite.)

While the region was very small, there was plenty of time for it to have come to a uniform temperature. So in the inflationary model, the uniform temperature was established before inflation took place, in a very, very small region. The process of inflation then stretched this very small region to become large enough to encompass the entire observed universe. Thus, the sources of the microwave background radiation arriving today from all directions in the sky were once in close contact; they had time to reach a common temperature before the inflationary era began.

The inflationary model also provides a simple resolution for the third question in table 1.1, the issue of the mass density. Recall that the ratio of the actual mass density to the critical density is called Ω and that the problem arose because the condition $\Omega = 1$ is unstable:

Ω is always driven away from 1 as the universe evolves, making it difficult to understand how its value today can be in the vicinity of 1.

During the inflationary era, however, the peculiar nature of the false vacuum state results in some important sign changes in the equations that describe the evolution of the universe. During this period, as we have seen, the force of gravity acts to accelerate the expansion of the universe rather than to retard it. It turns out that the equation governing the evolution of Ω also has a crucial change of sign: during the inflationary period the universe is driven very quickly and very powerfully *towards* a critical mass density.

So a very short period of inflation can drive the value of Ω very accurately to 1, no matter where it starts out. There is no longer any need to assume that the initial value of Ω was incredibly close to 1.

Furthermore, a prediction comes out of this. The mechanism that drives Ω to 1 almost always overshoots, which means that even today the mass density should be equal to the critical mass density to a high degree of accuracy. More precisely, the model predicts that the value of Ω today should equal 1 to an accuracy of about 1 part in 10,000. (The deviations from 1 are caused by quantum effects, which I will talk about shortly.) Thus, the determination of the mass density of the universe would be a very important test of the inflationary model.*

Unfortunately, it is very difficult to reliably estimate the mass density of the universe. Part of the reason is the fact that most of the mass in the universe is in the form of "dark matter," matter which is totally unobserved except for its gravitational effects on other forms of matter. Since we do not even know what the dark matter is, it is very difficult to estimate how much of it exists. Most of the current estimates, I must admit, give values for Ω that are distinctly below 1: numbers like 0.1 to 0.3 are most common. But these estimates are highly uncertain, and there appears to be no compelling observational evidence at present to rule out the possibility that $\Omega = 1$.

Finally, I come to the last of the four questions, concerning the origin of the primordial density perturbations in the universe. The generation of density perturbations in the new inflationary universe was addressed in the summer of 1982 at the Nuffield Workshop on the Very Early Universe, held at Cambridge University. A

---

*In the text I have followed the common assumption that Einstein's cosmological constant Λ is either zero or negligible. Otherwise the prediction becomes $\Omega + (\Lambda/3H^2) = 1$.

number of theorists were working on this problem, including Steinhardt, James M. Bardeen of the University of Washington, Stephen W. Hawking of Cambridge University, So-Young Pi of Boston University, Michael S. Turner of the University of Chicago, A. A. Starobinsky of the L. D. Landau Institute of Theoretical Physics in Moscow, and myself. It was found that the new inflationary model, unlike any previous cosmological model, leads to a definite prediction for the spectrum of perturbations. Basically the process of inflation first smooths out any primordial inhomogeneities that may have been present in the initial conditions. For example, any particles that may have been present before inflation are diluted to a negligible density. In addition, the primordial universe may have contained inhomogeneities in the gravitational field, which is described in general relativity in terms of bends and folds in the structure of spacetime. Inflation, however, stretches these bends and folds until they become imperceptible, just as the curvature of the surface of the earth is imperceptible in our everyday lives.

For a while, we were worried that inflation would give us a totally smooth universe, which would obviously be incompatible with observation. It was pointed out, however, I believe first by Hawking, that the situation might be saved by the application of quantum theory.

A very important property of quantum physics is that nothing is determined exactly—everything is probabilistic. Physicists are of course accustomed to the idea that quantum theory, with its probabilistic predictions, is essential to describe phenomena on the scale of atoms and molecules. On the scale of galaxies or clusters of galaxies, on the other hand, there is usually no need to consider the effects of quantum theory. But inflationary cosmology implies that for a short period the scales of distance increased very rapidly with time. Thus, the quantum effects which occurred on very small, particle-physics length scales were later stretched to the scales of galaxies and clusters of galaxies by the process of inflation.

Therefore, even though inflation would predict a completely uniform mass density by the rules of classical physics, the inherent probabilistic nature of quantum theory gives rise to small perturbations in the otherwise uniform mass density. The spectrum of these perturbations was first calculated during the exciting three-week period of the Nuffield workshop. After much disagreement and discussion, the various working groups came to an agreement on the answer. I will describe these results in two parts.

First of all, we calculated the shape of the spectrum of the perturbations. The concept of a spectrum of density perturbations may seem a bit foreign, so let me explain that the analogy of sound waves is very good. People familiar with acoustics understand that no matter how complicated a sound wave is, it is always possible to break it up into components which each have a standard wave form and a well-defined wavelength. The spectrum of the sound wave is then specified by the strength of each of these components. In discussing density perturbations in the universe, it is similarly useful to define a spectrum by breaking up the perturbations into components of well-defined wavelength.

For the inflationary model, we found that the predicted shape for the spectrum of density perturbations is essentially scale-invariant; that is, the magnitude of the perturbations is approximately equal on all length scales of astrophysical significance. While the precise shape of the spectrum depends on the details of the underlying grand unified theory, the approximate scale-invariance holds in almost all cases. It turns out that a scale-invariant spectrum was proposed in the early 1970s as a phenomenological model for galaxy formation by Edward R. Harrison of the University of Massachusetts at Amherst and Yakov B. Zel'dovich of the Institute of Physical Problems in Moscow, working independently.

Unfortunately, there is still no way of inferring the precise form of the primordial spectrum from observations, since one cannot reliably calculate how the universe evolved from the early period to the present. Such a calculation is very difficult in any case, and it is further complicated by the uncertainties about the nature of the dark matter. Nonetheless, the scale-invariant spectrum appears to be at least approximately what is needed to explain the evolution of galaxies, and thus this prediction of the inflationary model appears so far to be successful. Galaxy formation is currently a very active subject of research, so a better determination of the spectrum of primordial density perturbations may be developed. Such a result would provide an additional test of the inflationary universe model.

The predicted magnitude of the density perturbations was also calculated by the group at the Nuffield workshop, but the implications of these results were much less clear. It was found that the predicted magnitude, unlike the shape of the spectrum, is very sensitive to the details of the underlying particle theory. At the time the minimal $SU(5)$ theory, the first and simplest of the grand unified theories, was strongly favored by anybody interested in grand unified theories. We were therefore very disappointed when we found that the minimal $SU(5)$ theory leads to density perturbations with a magnitude that is 100,000 times larger than what is desired for the evolution of galaxies. Thus, there was a serious incompatibility between the inflationary model and the simplest of the grand unified theories.

With the passage of time, however, the credibility of the minimal $SU(5)$ grand unified theory has diminished. The minimal $SU(5)$ theory makes a rather definite prediction for the lifetime of a proton, and a variety of experiments have been set up to test this prediction by looking for proton decay. So far no such decays have been observed, and the experiments have pushed the limit on the proton lifetime to the point where the minimal $SU(5)$ theory is now excluded.

With the exclusion of the minimal $SU(5)$ theory, a wide range of grand unified theories become plausible. All of the allowed theories seem a bit complicated, so apparently we will need some kind of new understanding to choose which, if any, is the correct theory.

A variety of grand unified theories that predict an acceptable magnitude for both the proton lifetime and the density perturbations have been constructed. Thus, while the inflationary model cannot be credited with correctly predicting the magnitude of the perturbations, it also cannot be criticized for making a wrong prediction. The situation is very similar to the calculation of the net baryon number of the universe that I discussed earlier: the inflationary model provides at least a framework for calculating the magnitude of the density perturbations. If sometime in the future the correct grand unified theory and the values of its free parameters somehow become known, it will then be possible to make a real theoretical prediction for the magnitude of the perturbations.

A common feature of the known models that lead to acceptable density perturbations is the abandonment of the idea that inflation can be driven by the Higgs fields that break the grand unified symmetry. It appears that any Higgs field that interacts strongly enough to break the grand unified symmetry leads to density perturbations with a magnitude that is far too large. Thus, it must be assumed that the underlying particle theory contains a new field— a field which strongly resembles the Higgs fields in its properties, but which interacts much more weakly than the Higgs fields.

Unfortunately, all of the known theories that give acceptable predictions for the magnitude of the density perturbations look a little contrived. To be honest, the theories *were* contrived—with the goal of getting the density perturbations to come out right. The need for this contrivance can certainly be used as an argument against the inflationary model, but in my opinion this argument is considerably weaker than the arguments in favor of inflation. Even if we ignore cosmology, any grand unified theory that is consistent with the known properties of particle physics appears to be rather contrived. Clearly there are some fundamental principles at work here that we do not yet understand.

I would like to emphasize that my allusions to fundamental principles beyond the grand unified theories are not based on idle speculation—they are based on active and energetic speculation. Since the invention of grand unified theories in 1974, particle theorists have been vigorously working on attempts to construct the ultimate theory of nature—an elegant theory which would include a quantum description of gravity. The characteristic energy scale of such a theory is presumably the Planck scale, 10^{19} GeV, the scale at which the gravitational interactions of elementary particles become comparable in strength to the other types of interactions. It is then hoped that a grand unified theory would emerge as a low energy approximation.

The latest and most successful of these attempts is a radically new kind of particle theory known as a "superstring theory." Superstring theories represent a dramatic departure from conventional theories in that particles are viewed as ultramicroscopic strings (length $\approx 10^{-33}$ centimeter). Furthermore, according to the theory, the universe has nine spatial dimensions. Early in the history of the universe, when the temperature cooled below $10^{32}°$K, all spatial dimensions but the three we know today stopped expanding and remained curled up with an unobservably small radius. As bizarre as the theory may sound, the superstring theory has been shown to possess a number of unique properties crucial to a quantum theory of gravity, and it has totally captured the attention of a large fraction of the worldwide particle theory community.

At present very little can be said about the behavior of superstring theories at energies well below the Planck scale. Nonetheless, it is encouraging to know that progress is under way toward embedding the idea of grand unification into a larger framework. Superstring theories are highly constrained, which leads to hopes that someday we may be able to make rather definite predictions concerning physics at the energy scale of grand unified theories and beyond. If such a success is ever achieved, then the calculation of the predicted spectrum of density perturbations will provide a very rigorous test of inflationary cosmology.

Finally, I want to mention that quantum effects during the inflationary era are not the only source of primordial density perturbations that particle physics can provide. There is also the possibility, which I will not discuss in detail, that the seeds for galaxy formation may have been objects called "cosmic strings" (not related to superstrings). These strings are predicted to exist by some but not all grand unified theories, and they would form in a random pattern during the grand unified theory phase transition. As their name suggests, strings are very thin, spaghetti-like objects that can form infinite curves or closed loops of astrophysical size. A cosmic string has a thickness of about 10^{-29} centimeter and a mass of about 10^{22} grams for each centimeter of length. (In astronomical terms, the mass is about 10^7 solar masses per light-year.) In most theories the density of these strings would be diluted to negligibility by the process of inflation, but it is possible to construct theories in which the strings survive by forming either after inflation or at the very end of it. Cosmic strings are a very active topic of current research, and it is beginning to appear that a number of features of galactic structure can be explained naturally in terms of cosmic strings. Models of this type still make use of inflation to answer the second and third questions of table 1.1, and also to smooth out any small-scale inhomogeneities which may have been present in the initial conditions.

CONCLUSION

In summary, I feel that the inflationary universe model has been very successful in describing the broad, qualitative properties of the universe. In particular, the model provides very attractive answers to the four questions discussed in this chapter. While the model must be treated as speculative, I nonetheless feel that in its broad outline the inflationary universe model is essentially correct.

The inflationary model makes two observationally testable predictions: it predicts the mass density of the universe and also the shape of the spectrum of primordial density perturbations. While neither of these predictions is straightforward to check, it nonetheless seems likely that significant progress will be made in the foreseeable future.

Even if the inflationary model is correct, however, it must still be emphasized that nothing I have discussed is a completed project. The inflationary model is not a detailed theory; it is really just an outline for a theory. Michael Turner calls it the "inflationary paradigm" (Turner 1987). In order to fill in the details, we will need to know much more about the details of particle physics at the energy scales of grand unified theories and perhaps beyond.

So it appears to me that the fields of particle physics and cosmology will be closely linked for some years to come, as physicists continue their efforts to understand the fabric of space, the structure of matter, and the origin of it all.

References

Abbott, L. F. and S.-Y. Pi, eds. 1986. *Inflationary Cosmology*. Singapore: World Scientific.

Albrecht, A. and P. J. Steinhardt. 1982. Cosmology for grand unified theories with radiatively induced symmetry breaking. *Physical Review Letters* 48:1,220–23.

Bernstein, J. and G. Feinberg, eds. 1987. *Cosmological constants: Papers in modern cosmology*. New York: Columbia University Press.

Blau, S. K. and A. H. Guth. 1987. Inflationary cosmology. In *300 years of gravitation*. Ed. S. W. Hawking and W. Israel. Cambridge: Cambridge University Press.

Brandenberger, R. H. 1985. Quantum field theory methods and inflationary universe models. *Reviews of Modern Physics* 57:1–60.

Dicke, R. H. and P. J. E. Peebles. 1979. The big bang cosmology—enigmas and nostrums. In *General relativity: An Einstein centenary survey*. Ed. S. W. Hawking and W. Israel. Cambridge: Cambridge University Press.

Einstein, A. 1916. Die Grundlage der allgemeinen Relativitätstheorie. *Annalen der Physik* 49:769–822. English translation in Lorentz, Einstein, Minkowski, and Weyl 1952.

———. 1917. Kosmologische Betrachtungen zur allgemeinen Relativitätstheorie. *Sitzungsberichte der Preussischen Akademie der Wissenschaften* 1917:142–52. English translation in Lorentz, Einstein, Minkowski, and Weyl 1952 and in Bernstein and Feinberg 1987.

Gasiorowicz, S. and J. L. Rosner. 1981. Hadron spectra and quarks. *American Journal of Physics* 49:954–84.

Georgi, H. and S. L. Glashow. 1974. Unity of all elementary-particle forces. *Physical Review Letters* 32:438–41.

Georgi, H., H. R. Quinn, and S. Weinberg. 1974. Hierarchy of interactions in unified gauge theories. *Physical Review Letters* 33:451–54.

Gribbin, J. 1986. *In search of the big bang: Quantum physics and cosmology*. London: Heinemann.

Guth, A. H. 1981. Inflationary universe: A possible solution to the horizon and flatness problems. *Physical Review* D23:347–56.

Guth, A. H. and P. J. Steinhardt. 1987. The inflationary universe. To be published in *The new physics*. Ed. Paul Davies. Cambridge: Cambridge University Press.

Linde, A. D. 1982. A new inflationary universe scenario: A possible solution of the horizon, flatness, homogeneity, isotropy and primordial monopole problems. *Physics Letters* 108B:389–93.

———. 1984. The inflationary universe. *Reports on Progress in Physics* 47:925–85.

Lorentz, H. A., A. Einstein, H. Minkowski, and H. Weyl. 1952. *The principle of relativity: A collection of original memoirs on the special and general theory relativity*. Reprint, translated by W. Perrett and G. B. Jeffery. New York: Dover.

Penzias, A. A. and R. W. Wilson. 1965. A measurement of excess antenna temperature at 4080 Mc/s. *Astrophysical Journal* 142:414–19.

Sakharov, A. D. 1967. Violation of CP invariance, C asymmetry, and baryon asymmetry of the universe. *Soviet Physics, JETP Letters* 5:32–35.

Schramm, D. N. 1983. The early universe and high-energy physics. *Physics Today*, April 1983:27–33.

Silk, J. 1980. *The big bang: The creation and evolution of the universe*. New York: W. H. Freeman.

Steinhardt, P. J. 1986. Inflationary cosmology. In *High Energy Physics, 1985*. Proceedings of the Yale Theoretical Advanced Study Institute. Vol. 2. Ed. M. J. Bowick and F. Gürsey. Singapore: World Scientific.

Tryon, E. P. 1987. Cosmic inflation. In *The encyclopedia of physical science and technology*. Vol. 3. New York: Academic Press.

Turner, M. S. 1987. Cosmology and particle physics. In *Architecture of fundamental interactions at short distances*. Ed. P. Ramond and R. Stora. Amsterdam: North Holland.

Weinberg, S. 1977. *The first three minutes: A modern view of the origin of the universe*. New York: Basic Books.

———. 1979. Cosmological production of baryons. *Physical Review Letters* 42:850–53.

Yoshimura, M. 1978. Unified gauge theories and the baryon number of the universe. *Physical Review Letters* 41:281–84. Erratum ibid. 42:746.

GEORGE W. WETHERILL

2 Formation of the Earth

About ten years ago, while I was on a ride in what used to be called an amusement park, a disembodied head appeared above me out of the darkness. Its lips quivered as it intoned: "So you want to know how the world began? Very well then, I will show you. But remember, you fool, no one asked you to take this journey!"

Although I would not express myself in the dire tones employed by that specter, I feel that some kind of warning must be given to the reader at the outset of this presentation. You should know that there really are no bona fide experts on the formation of the Earth. While I am reluctant to confess to being an imposter, one must not forget that all the events I will be discussing took place a long time ago, and the vestigial observational record of them has been heavily veiled by the subsequent history of the Earth and solar system. I would be misleading you if I claimed there was any certainty that much of this will ultimately turn out to be really true. In spite of these reservations, I have become persuaded that a number of

The new work on the early stage of planetary accumulation described here was based on new expressions for velocity changes in a planetesimal swarm provided by G. R. Stewart. I wish to thank Janice Dunlap for her work in preparing this manuscript. This work was supported by NASA grant NSG–7347 and is part of a general departmental program supported by NASA grant NAGW–398.

worthwhile things can be said about how the Earth was formed 4.5 billion years ago, that the number of these things has increased quite a bit in recent years, that at least some ways of looking at these problems are much more likely to be correct than others, and that there is a story worth telling, even though it could turn out to be a fairy tale.

At one time, and to some extent even today, theories of the formation of the Earth and solar system tended to be very idiosyncratic; they were associated with the names of particular individuals who attempted all by themselves to present a complete explanation of solar system formation. For the most part these theories did not intersect one another. Now this field is evolving into a more normal science, in which one can feel rewarded for contributing to the solution of relevant but finite problems, and in which it is becoming more customary to build upon the work of one's predecessors and colleagues, rather than always treating them as adversaries.

At least with regard to the physical aspects of this problem, much of the credit for this change must be assigned to a group of scientists in the Institute of the Physics of the Earth in Moscow—Victor Safronov, his wife Evgenia Ruskol, and a few younger collaborators. This work first came to the attention of most Western scientists in the early 1970s, following NASA's translation into English of Safronov's *Evolution of the Protoplanetary Cloud and Formation of the Earth and the Planets* (Safronov 1969). This book, summarizing the work done by Safronov and his colleagues during the previous 15 years, posed a number of well-defined problems that can be addressed one at a time. Although this work made important contributions to our understanding of all of these problems, it would be too much to expect that definitive solutions to any of them could be obtained so early. Probably the most important aspect of Safronov's work was the laying out of a systematic program of investigations that gives those of us working in this field reason to believe we are climbing a ladder, rather than running on a treadmill.

As a result, there has emerged what can be considered a "standard model" for the formation of the solar system, and of the Earth and other terrestrial planets in particular. This model is of course unlikely to be true in detail and may even prove to be qualitatively incorrect. Nevertheless, it has had some encouraging successes and provides a disciplined framework for developing an internally con-

TABLE 2.1 "Standard model" for the formation of the terrestrial
planets by the accumulation of smaller planetesimals

---

1. The sun formed from an interstellar molecular cloud, requiring a time of $\sim 10^5$ years.

2. Planetesimals (about 10 km in diameter) formed in nebula of residual dust and gas left over from the formation of the sun. Time scale $\sim 10^4$ years.

3. Planetesimals grew by colliding with one another at velocities about equal to their escape velocities. Growth of planetesimals can be considered in two stages:

   A. Early "close-packed" stage. Bodies accumulated up to approximately lunar-size objects in concentric accumulation zones. Time scale: $10^5 - 10^6$ years. The size distribution of these bodies depends on the extent to which "runaway" growth occurred during this stage.

   B. The large planetesimals that formed in stage A collided and merged to form the final planets. During this growth it is likely that wide-spread mixing of bodies occurred throughout the entire terrestrial planet region. Time scale: $10^7 - 10^8$ years.

   A dynamically significant quantity of nebular gas was present at least during the beginning of stage A, and may have been present later as well.

---

sistent explanation for the formation of the terrestrial planets: Earth, Venus, Mars, and Mercury. I will talk primarily about my efforts, building upon this work of others, to advance our understanding of problems associated with this standard model.

What is the standard model? As outlined in table 2.1, it starts with the dense molecular clouds ($\sim 10^5$ hydrogen atoms/cm^3) observed by astronomers in regions of active star formation. As a by-product of the formation of the sun by the gravitational collapse and fragmentation of such a cloud, an initial state for the formation of the planets was achieved. Exactly how this evolution from a dense molecular cloud into a star surrounded by a far more compact disk of residual dust and gas took place is still poorly understood, but it is an active and exciting field of theoretical and observational astronomy (see Boss 1985).

The distribution of dust and gas (commonly referred to as the "solar nebula") envisioned as the initial state of the standard model is shown in cartoon form in figure 2.1. At the center is the sun, with

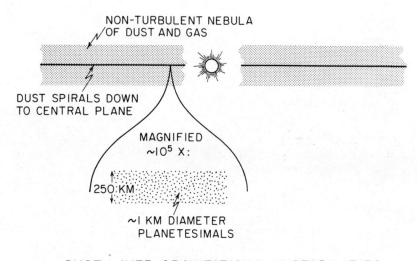

NON-TURBULENT NEBULA
OF DUST AND GAS

DUST SPIRALS DOWN
TO CENTRAL PLANE

MAGNIFIED
~10^5 X:

250 KM

~1 KM DIAMETER
PLANETESIMALS

DUST LAYER GRAVITATIONAL INSTABILITIES

Fig. 2.1. *Schematic representation of the solar nebula (the residual gas and dust left over from the formation of the sun). This nebula was probably highly turbulent initially (in fact, turbulence may have been necessary in order to provide it with the angular momentum necessary to prevent it from falling into the sun). As turbulence decreased, the dust component settled down to form a thin layer (about 250 km thick) in the central plane of the nebula. Planetesimals of about 10 km in diameter formed, possibly by gravitational instabilities in the dust layer.*

nearly its present mass, but not yet on the "main sequence" of stellar evolution. Surrounding the sun is a flattened disk of dust and gas. This disk was probably turbulent at one time. As the turbulence decayed away, the dust separated from the gas. As a result of "gas drag"—slowing down of the dust particles by friction as they move through the gas—these particles would have spiraled down to the central plane of the disk and formed the thin, dust-rich layer shown by the heavy line in figure 2.1.

An important possible consequence of this concentration of dust in the central plane was first described by Edgeworth (1949) and subsequently discussed further by Safronov and a number of other workers. This consequence is that the self-gravity of a sufficiently nonturbulent and dense central dust layer may cause it to spontaneously break up into a large number of "planetesimals" about 1 to 10 kilometers in diameter. Whether a real solar nebula is sufficiently nonturbulent to do this is among the questions posed by the model. The answer is not yet known. If this theory does not work, there are other possible ways to form initial planetesimals—for example, clumping as a result of the "stickiness" of the dust particles (Weidenschilling 1980).

Regardless of how these planetesimals actually formed, the further evolution of the standard model involves understanding how they became aggregated into larger bodies and ultimately into planets. The present system of terrestrial planets has a mass of about 10^{28}g. Therefore, at least about 10^{10} bodies, each about 10^{18}g in mass (10 kilometers in diameter), must be considered. At first, these bodies will be spread out in nearly circular orbits in the general region of heliocentric distance occupied by the present terrestrial planets. The problem at hand is understanding the manner in which a well-defined physical system of this kind may be expected to evolve naturally, without our sticking our fingers in it and saying "I want you to do this" or "I want that to happen." Under what circumstances will this system of small planetesimals evolve into a system that resembles our present system of terrestrial planets, and under what circumstances will it evolve in some different way?

A basic requirement for evolution into larger planets is that when these planetesimals collide with one another, at least most of the time they merge and thereby grow larger. Whether this will happen depends on the relative velocity of the two bodies. If they are of more or less similar mass and collide with velocities well above their "escape velocity," then a large fraction of the collision fragments will acquire velocities exceeding the escape velocity and the bodies cannot grow. The ratio of the relative velocities of the bodies to their escape velocity is therefore a crucial parameter.

Many early treatments of the problem of planetesimal growth treated this velocity ratio as a free parameter; the fundamental question of whether growth was possible was not constrained by the theory. If the velocities of the planetesimals result from their mutual interactions rather than external forces, then in fact the relative velocity is not a free parameter; rather, it is determined by the balance between the increase of relative velocity caused by mutual gravitational perturbations of the bodies and the loss of velocity caused by their collisions. One of Safronov's principal contributions was to devise a theory that permits calculation of the relative velocity of an ensemble of bodies of given sizes moving in heliocentric orbits. He found that the gravitational perturbations and collisions balanced one another in such a way that the resulting relative velocity turns out to be comparable to or less than the escape velocity of the larger bodies of the ensemble. Thus, these larger bodies can grow by sweeping up smaller bodies, rather than being destroyed by fragmentation when these collisions occur.

Safronov's result was obtained by use of an analytic theory in which the planetesimals were considered as analogous to gas molecules in the kinetic theory of gases. A complementary approach is a numerical one, in which the fact that the planetesimals are actually constrained to move in heliocentric orbits is explicitly introduced.

The results of several such numerical calculations are shown in figure 2.2 (Wetherill 1980a, b). The spontaneous evolution of a swarm of a hundred equal-mass bodies moving in heliocentric orbits was simulated by a computer calculation. When any two of these hypothetical bodies approached each other closely, the change in their orbits resulting from their mutual gravitational interaction was calculated. When they came close enough to collide, their relative velocity was reduced to nearly zero, but in the "pure" case considered in this calculation they did not merge to form a larger body. In three of these calculations the initial relative velocities of the bodies were chosen to be much lower than their escape velocity of 1.2 km/sec. Three of the four symbols—solid squares, open circles, and open squares—represent three repeats of essentially the same calculations. (They differ slightly from one another because the stochastic nature of the encounter process is simulated by use of random numbers in the so-called Monte Carlo method.) In these three cases, the relative velocities at first rose rapidly as their mutual gravitational perturbations "pumped up" their eccentricities and inclination. This can be thought of as a random version of the use of "gravity assist" by interplanetary spacecraft, whereby high velocity orbits are achieved by deliberately flying close to a planet or the moon. As the velocities increased, the energy loss caused by their collisions became more important, causing the relative velocity to increase less rapidly. Ultimately, the effects of gravitational perturbations and collisions came into balance and an approximately steady-state relative velocity of about 1.5 km/sec, only slightly higher than the 1.2 km/sec escape velocity, was reached. This result agrees closely with Safronov's analytical theory. That this is a true equilibrium is demonstrated by a fourth calculation (solid circles) in which the initial

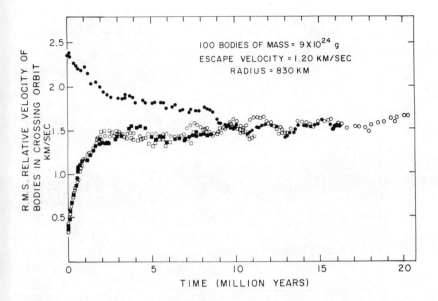

Fig. 2.2. *Numerical calculations of the approach to a steady-state mean velocity of a nonaccumulating swarm of bodies of equal mass. The equilibrium velocity is found to be nearly independent of the initial velocity. (The time scale shown applies to a swarm of bodies of the size indicated and has no direct relevance to the time scale for formation of the actual solar system.)*

velocity was chosen to be well above 1.5 km/sec. In this case the effect of collisions is predominant at the outset, and with the passage of time the relative velocity decreases to the same steady-state value. (The time scale is determined by the number and size of the bodies chosen and has no direct relevance to the time scale for solar system formation.)

This same calculation has been repeated for bodies of different initial sizes. In every case a steady-state velocity is achieved. The manner in which these steady-state velocities vary with the radius and escape velocity of the body is shown in figure 2.3. The relationship is found to be linear. As the bodies grow larger, the ratio of relative velocity to escape velocity remains constant, permitting growth to continue. As mentioned earlier, identification of this relationship was a major achievement of Safronov's investigations. Of course, the situation will be more complex in a real swarm that contains bodies of varying sizes. But the result remains the same:

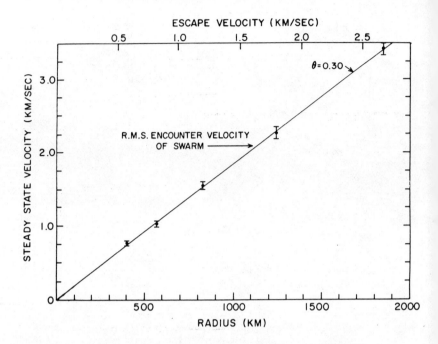

Fig. 2.3. *Steady-state velocities found by numerical calculations of nonaccumulating swarms of bodies of equal size, as a function of their radius. The relative velocities are found to increase linearly with the radius and escape velocity, in agreement with Safronov's analytical theory.*

the relative velocities of the bodies that contain most of the mass of the swarm are constrained to be similar to or less than the escape velocity of the larger bodies.

FIRST STAGE OF PLANETESIMAL ACCUMULATION

As stated in table 2.1, following the formation of the planetesimals it is convenient to divide the evolution of the standard model into two consecutive stages of planetesimal growth. During the first stage, the number of bodies is very large, of the order of 10^{10}, and it is completely out of the question to calculate and follow the orbits of each individual planetesimal. Instead, it is necessary, and reasonably correct, to treat the swarm not as a number of individual particles by the methods of orbital dynamics, but in a way similar to the treatment of stars in Chandrasekhar's stellar dynamic theory (Chandrasekhar 1942)—that is, as analogous to molecules in molecular theories of gases. The first stage is illustrated schematically

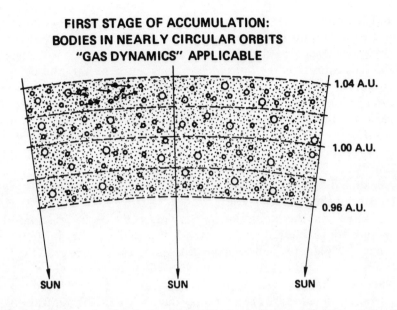

FIRST STAGE OF ACCUMULATION:
BODIES IN NEARLY CIRCULAR ORBITS
"GAS DYNAMICS" APPLICABLE

1.04 A.U.

1.00 A.U.

0.96 A.U.

SUN   SUN   SUN

Fig. 2.4. *Schematic representation of planetesimals during the first stage of planetesimal growth. The large number of small planetesimals will interact with one another locally within concentric heliocentric zones about 0.02 AU in width. The evolution of their velocity and mass distributions can be considered by the techniques of gas dynamics.*

in figure 2.4. We can conceptually divide the region in which the terrestrial planets are to form into concentric bands about 0.02 AU in width, extending from inside the present orbit of Venus to beyond the present orbit of the Earth. Within each of these zones one can consider the relative motion of the planetesimals in a reference frame moving at the local Keplerian velocity about the sun. As a consequence of the small but nonzero values of their eccentricities and inclinations, the bodies will have slightly different orbits. Orbits of bodies will cross one another, and when collisions occur at orbital intersections, the bodies will have a significant relative velocity with respect to one another. During this initial stage, the objects in each zone will be "ignorant" of what is going on in adjacent zones. In the rotating reference frame of each of the zones, the motion approximates the motion familiar from the kinetic theory of gases.

Safronov and his colleagues calculated the motion by using approximate expressions ultimately based on Chandrasekhar's work in stellar dynamics. More recently, Stewart and Kaula (1980) and Hornung, Pellat, and Barge (1985) have made important contributions to this discussion. In particular, Stewart and Wetherill (1988) have provided quite general expressions for the velocity changes caused by the mutual collisions and gravitational perturbations of the planetesimals in a form that is useful for numerically calculating the simultaneous velocity and mass evolution of a planetesimal swarm. One of these terms, *dynamical friction*, was not included at all in the earlier work of Safronov and others. The most important consequence of this term is that it tends to "equipartition" kinetic energy between bodies of different sizes, thereby decreasing the velocity of the larger bodies and increasing the velocities of the smaller bodies.

I have used the expressions of Stewart as the basis for a numerical procedure that follows the growth and velocity distribution of a swarm of planetesimals during this early stage of a growth. An initial distribution is assumed in which all of the bodies are concentrated in a narrow size range corresponding to the ~10 km-diameter bodies formed by gravitational instability in the central dust layer of figure 2.1. The initial velocity is taken to be equal to the escape velocity of bodies of a size such that half of the mass of the ensemble is in larger bodies and half in smaller bodies. The actual continuous initial mass distribution is approximated by dividing it into a number (about 10) of "batches" of the same mass, each having its own characteristic velocity relative to the Keplerian circular velocity at a heliocentric distance of 1 AU—that is, at the orbit of the Earth. During an initial time interval, the change of velocity of the bodies in each of the batches as a result of their mutual gravitational perturbations and collisions is calculated. Growth of the mass of the bodies in each batch is included by calculating the number of bodies of equal or smaller size that are swept up during the time interval. Repetition of this calculation during successive time intervals permits calculation of the evolution of the size and velocity distribution with the passage of time. In addition to the simple sweep up of considerably smaller bodies, collisions between bodies of comparable masses will form bodies that are more massive than the mean mass of the largest batch. When this occurs, new batches are created. In this way, the dispersion of the mass distribution within the batches is included in an approximate way.

The results of one calculation are shown in figure 2.5. This calculation is of interest because it numerically duplicates the conditions of the problem for which Safronov has obtained an analyt-

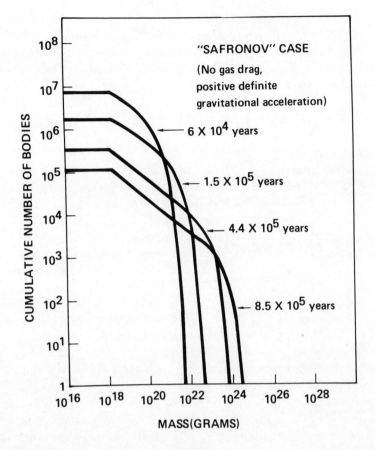

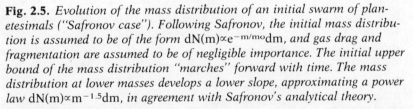

Fig. 2.5. *Evolution of the mass distribution of an initial swarm of planetesimals ("Safronov case"). Following Safronov, the initial mass distribution is assumed to be of the form dN(m)∝e$^{-m/m_0}$dm, and gas drag and fragmentation are assumed to be of negligible importance. The initial upper bound of the mass distribution "marches" forward with time. The mass distribution at lower masses develops a lower slope, approximating a power law dN(m)∝m$^{-1.5}$dm, in agreement with Safronov's analytical theory.*

ical solution, that is, it ignores the effect of dynamical friction. Although the expression derived by Stewart for the gravitational "pumping" of the relative velocities is included, the principal features of the Safronov result are preserved. An initial mass distribution corresponding to that used by Safronov is given by the curve marked $T = 0$. Again, following Safronov, the effects of "gas drag" are assumed to be unimportant in this idealized calculation. From figure 2.5 it is seen that, as time goes by, the upper bound of the mass distribution migrates forward in mass, while at the same time the number of bodies in this "marching front" becomes smaller. At the small-mass end of the distribution, the number of bodies dn between m and a slightly larger mass $(m + dm)$ can be approximated by a power law such that

$$\frac{dn}{dm} = Cm^{-1.5}$$

where C is a constant determined simply by the size of the swarm. The calculation was continued until the increase in the relative velocities caused the eccentricities of the smaller bodies to increase to about 0.02, because the approximation of their having a uniform Keplerian velocity gradually introduces complications as the range of heliocentric distance increases. The result is somewhat similar to the result found analytically by Safronov: a power law with the same exponential index (-1.5), truncated by a marching upper bound.

Another opportunity to compare our approach to this problem with previous work is provided by including the effects of "gas drag," the slowing down of the bodies by friction with gas in the solar nebula. Gas (primarily hydrogen) must have been present at least at the earliest stage of planetary growth, because it is required to supply the material that formed the sun and the giant planets, Jupiter and Saturn. Accumulation in the presence of gas has been studied by C. Hayashi and his colleagues (Nakagawa, Hayashi, and Nakazawa 1983), again by using more restrictive assumptions. The results of our calculation are shown in figure 2.6. In both the gas-free case studied by Safronov and the gas-rich case studied by Hayashi and his colleagues, after about one million years much of the initial mass of the swarm is concentrated in a few bodies about 10^{25}g in mass, in agreement with the earlier work. This number of largest bodies corresponds to only one of the zones 0.02 AU in width depicted in figure 2.4. The total number in the terrestrial planet region in which most of the mass occurs in a zone about 0.4 AU in width will therefore be of the order of 10 to 100. A very similar result is found for the less idealized case where both gas drag and collisional damping are included. For the mass range in which most of the mass is concentrated, the effects of gas drag and collisional deceleration of planetesimal velocities are always found to be of comparable importance. It is not necessary to continue this kinetic-theory-of-gases-type of calculation beyond the stage at which most of the mass is concentrated in $\sim 10^{25}$g bodies, because the total

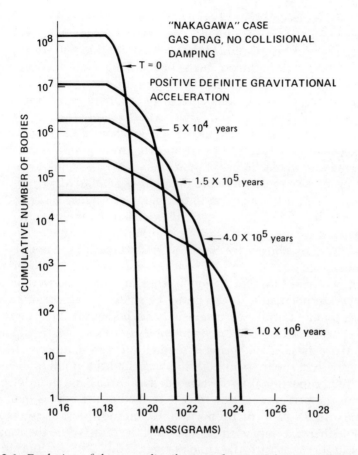

Fig. 2.6. *Evolution of the mass distribution of an initial swarm of planetesimals ("Nakagawa case"). Following Hayashi and his coworkers, the effects of gas drag are included, but collisional damping and fragmentation are not. The resulting distributions are quite similar to those formed for the Safronov case.*

number of large bodies is now in the range where it is possible to consider the evolution of individual orbits.

In the calculations shown in figures 2.5 and 2.6, as well as in the calculations of Safronov and Hayashi, the process of fragmentation has not been included. Although the larger bodies of the swarm that comprise the "marching front" of figure 2.5 are big enough to be immune to significant fragmentation, the relative velocities of the smaller bodies reach values large enough to cause them to be totally fractured. Qualitatively, this will have the effect of causing the smaller end of the size spectrum to approach a quasi-steady-state balance between input of collision fragments of somewhat larger bodies and loss by fragmentation. At the same time, the bodies comprising the upper end of the size distribution will simply grow. It is not difficult to introduce the phenomena of fragmentation into these calculations in a formal way, and this has been done. At the same time, it must be recognized that realistic modeling of such massive collisions between large bodies severely taxes our present understanding of the fragmentation mechanics of material even when its physical properties are known. Any assumptions about the fragmentation mechanics of these planetesimals of unknown physical properties must therefore be highly speculative.

Although it is satisfying to be able to reproduce independently the results of previous workers, extension of their work to less idealized assumptions suggests a different outcome. The initial distributions chosen by Safronov and Hayashi are not the only conceivable ones, and there is at present very little theoretical or observational basis for believing they are more likely than some others. In particular, it can easily be shown by algebraic calculations that an initial swarm consisting of bodies all of the same size, except for one body that is significantly (for example, by a factor of 30) more massive will grow in a quite different manner. If the rate of growth of the mass of both the large and the small bodies were simply proportional to their geometrical areas and otherwise uniform, the percentage growth rate of each of the bodies in a given time would *decrease* in proportion to their radii, because their areas would increase only with the square of the radii, while their masses would increase with the cube of their radii. Because its radius is larger, this decrease in percentage growth rate will be greater for the larger body than for the smaller one. As a result, when the growth rates are geometrical, the mass of the smaller body will tend to "catch up" with that of the larger body, in the same way that the relative value of a smaller bank deposit at a high rate of interest will catch up with that of a larger deposit at a lower rate of interest.

The growth rates will not be geometrical, however. Within a zone of the kind shown in figure 2.4, a larger body will have a tendency to grow more rapidly than the smaller bodies, because its stronger gravitational field will perturb the trajectories of bodies approaching it and tend to focus them onto a collision course. For this reason, its effective "gravitational radius" and corresponding "gravitational collision cross-section" will be effectively larger than its physical radius and cross-section. If this larger body is only a few times larger than the smaller body, this enhancement may be sufficient to cause its cross-section to be proportional to the cube of its radius,—that is, to its mass. In this case the tendency to "catch up" will disappear but the ratio of the masses of the large and small bodies will still tend to remain constant. This case was studied by Safronov 26 years ago by using an integro-differential coagulation equation (Safronov 1962).

If the larger body is sufficiently larger than the smaller one, however, the dependence of gravitational focusing on the radius will be greater than the cube of the radius. The result will be "runaway growth" of the larger body. This advantage of the larger body over the smaller bodies will continue to accelerate, limited only by the availability of smaller bodies to be swept up. The possible importance of early runaway growth of this kind preempting the evolution of the Safronov power-law size distribution was brought to general attention by numerical work of Greenberg, Wacker, Chapman, and Hartmann (1978).

When runaways occur, at first the size distribution is similar to those shown in figures 2.5 and 2.6. An important difference is the slope of the upper end of the mass distribution is seen to decrease with time, and at the end of the calculation the largest body has a mass several times that of the second largest body. This differs from the work of Safronov and Hayashi, who found a steep "marching front." The reason for this difference is the inclusion of the new term in the expressions of Stewart for the change of velocities of the bodies caused by their mutual gravitational perturbations. The effect of this term is to drive the velocity distribution of the swarm in the direction of "equipartition of energy." The more massive bodies will therefore tend to have lower velocities and the smaller bodies higher velocities. This has the effect of causing gravitational focusing to be greater than that considered by the earlier calculations in which these terms were not included. Although modeling of fragmentation is difficult, it appears likely that, in a rather complex way, the effects of fragmentation will also lead to a more rapid growth of the larger bodies.

Runaway growth can be demonstrated most simply by taking the same approach as that used to obtain figure 2.5, including an additional "seed" of mass 10 times larger than the second largest body and 170 times the mass of the smallest abundant mass in the swarm. The results are shown in figure 2.7. After 0.58 million years, a 10^{26}g body, more massive than the moon, has gathered up a large fraction of the mass originally contained in its zone. The rest of the

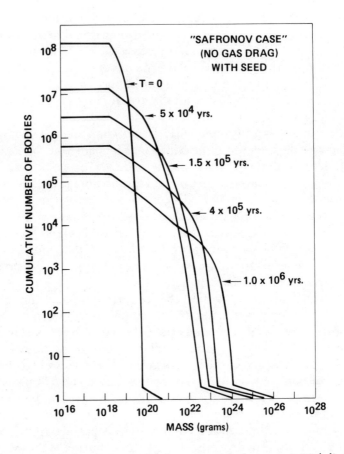

Fig. 2.7. *Evolution of planetesimal swarm with the same initial distribution as that of figures 2.5 and 2.6, except that a "seed" with a mass 10 times that of the second largest body is introduced into the initial swarm. This leads to a runaway, whereby the larger body grows to 10^{26}g (larger than the moon) and the second largest body only to 10^{24}g (less than the mass of the largest asteroid, Ceres).*

mass is concentrated in smaller bodies ($<10^{23}$g, or 370 km in radius) that have no chance of ever competing with the growth of the large body. Their ultimate fate can only be capture by the large body, or by similar large bodies in adjacent zones, or collisional fragmentation. The effect of including dynamical friction is similar to that of planting a "seed" in the original distribution. The lower velocity of the largest bodies allows the largest body to grow faster and the planetesimal swarm thereby "grows its own seed."

Despite representing a significant advance in theory over expressions used by previous workers, the expressions of Stewart used in this runaway calculation are not strictly applicable once the runaway is fully under way. For this reason, the details of the calculation cannot be interpreted exactly. At least semiquantitatively, however, growth of the sort shown in figure 2.7 must occur. The increase in collision cross-section that is responsible for the runaway can be shown to be a more general phenomenon than the strict applicability of Stewart's expressions, and it is this increase in cross-section that is ultimately responsible for the runaway. In fact, it is found that corrections should be applied to the theory in the runaway regime that work in the direction of further facilitating the runaway (Wetherill and Cox 1984, 1985). The principal uncertainty resulting from the failure of the theoretical expressions concerns the velocity and eccentricity distribution of the smaller bodies during the runaway. Formally, these eccentricities are found to be large enough to permit the smaller bodies to escape their zone of width 0.02 AU when the large body reaches a mass of about 10^{26}g. There are reasons to suspect that when the effects of fragmentation are included the large body may be able to sweep up a larger fraction of the 6×10^{26}g of the smaller bodies in its zone before this "escape" occurs. This is a problem of some importance that requires further study.

If a runaway occurs, how large can this body grow before the runaway is terminated by depletion of available material to sweep up? Quite early in the runaway the orbit of the large body is found to become nearly circular. The extent of the zone that can be swept clear by a body in an orbit of this kind has been studied by a number of workers, and is found to be about 10^3 times its physical radius, that is, ~0.01 AU for a 2,000 km-diameter body. Therefore, it seems plausible that a body of the final size shown in figure 2.7 could sweep up virtually all the material in its initial zone. At this size it seems likely that the runaway will slow down and possibly cease, either because similar runaways in adjacent zones have removed all the small material from the swarm, or because the large body is unable to perturb distant smaller bodies that may still be present into collision orbits.

In view of the foregoing rather qualitative discussion, it seems likely that in the runaway case, the outcome of the first "kinetic theory" stage of growth will be the concentration of most of the mass of the terrestrial planet region into a smaller number (about 20 to 40) of larger objects about the size of Mercury, rather than the about one thousand 10^{24} to 10^{25}g bodies found in the absence of a runaway. It should be clear to the reader that this discussion of the termination of a runaway has departed from the goal stated earlier, that of defining the initial conditions, letting the bodies do what comes naturally, and then "observing" the outcome. This remains to be done.

In view of the ease with which a runaway can occur, it would be tempting to assert that this is now known to be the outcome of the initial stage of terrestrial planet accumulation. However, such a firm conclusion, although probably correct, should be considered somewhat premature. In all these calculations it has been assumed that the relative velocities of the planetesimals were entirely determined by their mutual collisions and gravitational perturbations. This approach has the advantage of constituting a well-defined physical problem that should be studied in any case, if only as one building block that will be useful while considering more general models of planetary growth. On the other hand, it is possible that distant perturbations also played a role in determining the relative velocities of the planetesimals in the terrestrial planet region. For example, there are good (but not compelling) reasons to believe that Jupiter and Saturn were formed earlier than the terrestrial planets. If so, these massive planets and their protoplanetary forerunners may be expected to cause significant gravitational perturbations throughout the entire solar system. It is even conceivable that the relatively unstructured solar nebula may have influenced the eccentricities of the terrestrial planetesimals. At the present time, the giant planets generate eccentricities that can become as large as 0.07 in the orbits of Earth and Venus. Because the analogous eccentricities of the planetesimals would be to some extent in phase with one another, this would tend to reduce the effect of these distant perturbations on their relative velocities. For this reason, it is unlikely that these distant perturbations would have a major effect in the terrestrial planet region. Quantitative evaluation of the possible effect of distant perturbations in quenching runaways in the asteroid belt may be of greater importance and requires serious study.

FINAL STAGES OF ACCUMULATION

When the number of bodies containing the largest part of the mass becomes about 1,000 or smaller, it begins to be feasible to avoid problems associated with the kinetic-theory-of-gas approach and follow the natural orbital evolution of individual bodies. This liberates the theory from the restrictions imposed by ignoring the helio-centric spacing of the bodies and permits us to address questions relevant to the formation of the system of terrestrial planets, not just the growth of a single planet. Initially, the bodies will be localized in the vicinity of the zone in which they were formed, but on a fairly short ($\sim 10^5$ to 10^6 years) time scale it is found that their eccentricities increase and their semimajor axes migrate suffi-ciently to permit their range of interaction to include a large frac-tion of the terrestrial planet region. The result is a much looser aggregation of larger growing bodies, in contrast to the closely-packed situation shown in figure 2.4. During this earlier closely-packed stage, there was never a significant question of the larger bodies being able to sweep up smaller bodies in their zone. In the absence of runaway, each zone contained a number of bodies at the high-mass end of the Safronov "marching front" (figure 2.5) and these had ready access to the many relatively high-eccentricity smaller objects from which they were growing, the orbits of these smaller bodies always spanning those of several larger bodies. In the alternative case of runaway growth, the enhanced gravitational radius of the single large body was sufficient to sweep up its zone.

In contrast, during the final stage wherein 10^{25} to 10^{26} bodies are accumulated into the final terrestrial planets, the question of whether the orbits of the bodies permit collisions between them is of major importance. It ultimately determines whether the calcula-tions lead to the small number of terrestrial planets that are actu-ally observed, or to too many planets that are too small. Avoidance of the latter outcome requires that during the orbital evolution the mutual gravitational perturbations be sufficiently vigorous to maintain high enough relative velocities and hence eccentricities. On the other hand, if the velocities and eccentricities are too high, even bodies as large as the 10^{25} to 10^{26}g bodies under consideration will fragment rather than grow when they collide.

This is illustrated schematically in figures 2.8 and 2.9*a* and *b*. Figure 2.8 depicts the general situation during the final stage of growth; the orbits of only a few of the several hundred bodies are shown. The point being made is that during this stage, collision and growth require that the orbits of the bodies cross one another. This can be achieved in two ways: either the orbits are sufficiently eccen-tric or variations in the semimajor axes of the bodies cause the orbits to migrate radially. These two effects are found to be of comparable importance. Figure 2.9*a* illustrates the case where the eccentricities and relative velocities are low. In this case there is no

SECOND STAGE OF ACCUMULATION:
ECCENTRICITY AND/OR RADIAL MIGRATION
OF LARGE BODIES REQUIRED TO CONTINUE
GROWTH

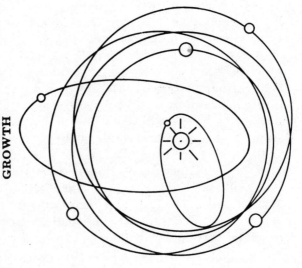

Fig. 2.8. Schematic representation of conditions controlling orbital evolution during the second stage of planetesimal evolution. Further growth toward planets of the size observed results in part by the low eccentricities of the lunar-size planetesimals formed in the first stage being "pumped up" to values high enough to permit them to collide with one another throughout the second stage of accumulation. Radial migration of semimajor axes is of comparable importance in maintaining the conditions for orbital intersection and planetary growth.

B. RELATIVE VELOCITY TOO HIGH

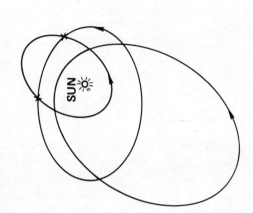

**HIGHLY ECCENTRIC ORBITS GIVE
HIGH VELOCITY ENCOUNTERS; PLENTY
OF COLLISION PARTNERS, BUT OBJECTS
FRAGMENT WHEN THEY COLLIDE AND
DON'T GROW.**

A. RELATIVE VELOCITY TOO LOW

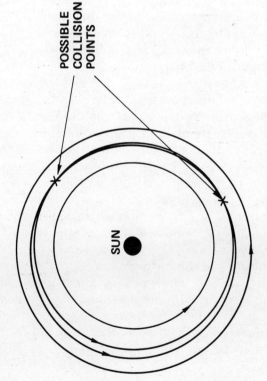

POSSIBLE
COLLISION
POINTS

**NEARLY CIRCULAR ORBITS GIVE LOW
VELOCITY ENCOUNTERS, BUT MOST
ORBITS DON'T INTERSECT.**

Fig. 2.9. *Schematic representation of conditions controlling orbital evolution during the second stage of planetesimal evolution. A: When the relative velocities are very low and the orbits nearly circular, collisional fragmentation is negligible and colliding bodies will merge and grow. These same nearly circular orbits preclude the bodies from having many collisional partners and favor evolution into a system of terrestrial planets different from that observed—that is, one consisting of a larger number of planets with smaller masses. B: When the relative velocities of the planetesimals are very high, there are plenty of collision partners, but the collisions are so energetic that the bodies will fragment rather than grow.*

problem with fragmentation, but most of the orbits fail to intersect one another. The number of collision partners is small and it is likely that the outcome of a situation of this kind will be too many planets that are too small. In figure 2.9*b*, the eccentricities and relative velocities are so high that there is no problem with collision partners, but the existence of the bodies is threatened by their high collisional velocities, which cause them to fragment.

Clearly, what is required is a velocity and eccentricity distribution that is neither too eccentric nor too circular. This is achieved naturally as a consequence of Safronov's result, illustrated in figures 2.2 and 2.3. The average collision velocity between the larger and smaller bodies is self-regulated to be comparable to the escape velocity of the larger, growing bodies. For bodies of the size of Earth and Venus, this corresponds to escape velocities of about 10 km/sec, implying eccentricities of 0.3. Smaller bodies with eccentricities this large at 1 AU will span a heliocentric distance from 0.7 to 1.3 AU, a range comparable to the dimensions of the entire terrestrial planet region.

The nature of the growth and orbital evolution that may be expected during the final stage of growth has been studied by means of Monte Carlo simulation (Wetherill 1980a, 1981, 1985a, 1986). In these calculations it is assumed that enough of the nebular gas has been swept up to cause it to be dynamically insignificant. The starting point is an assemblage of bodies with a size and velocity distribution of the kind found at the end of the earlier "kinetic theory" stage of growth (figures 2.5, 2.6, and 2.7). Calculations of this kind have been carried out for a variety of initial states in order to evaluate the sensitivity of the final result to the choice of initial state. One initial distribution that has been studied in some detail is shown in figure 2.10. Five hundred initial bodies are considered. These range in mass from 5×10^{24}g (about twice the mass of the largest asteroid, Ceres) up to 1.1×10^{26}g (a bit less than twice the mass of the moon), following a power law with an index of $-1\frac{1}{6}$ for the differential mass spectrum. Another initial distribution that has been studied in the same detail is one in which all 500 bodies are initially of equal mass. It is found that this system evolves rather quickly into one similar to that of figure 2.10, and the final results are similar. The same result is found for simulations in which the initial distributions correspond to local runaways in zones about 0.02 AU in width.

A few words should be said about the method of calculation. It might seem that the straightforward approach to this problem

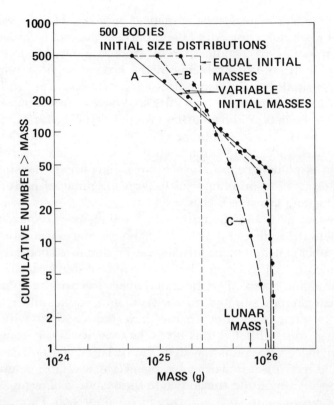

Fig. 2.10. *Initial distribution of 500 bodies (curve A). The steep upper bound represents the largest bodies formed in each of the concentric zones during the first stage of accumulation. The power law extending to smaller masses includes as many smaller bodies as can be accommodated within a swarm limited to 500 bodies. The other curves represent some of the alternative initial distributions that have been studied.*

might be to start with the initial distribution and integrate the three-dimensional Newtonian equations of motion forward in time for all 500 bodies. Upon more careful consideration, this turns out to be an absolutely impossible approach. Three-dimensional numerical integration of the accuracy required to distinguish between collisions and "near misses" of a 500-body problem for the required ~10^7 orbital periods goes far beyond any celestial mechanical calculation ever considered. Some alternative method must be found.

The alternative is based on a statistically valid method developed by E. J. Öpik (1951) to calculate the orbital evolution of planet-crossing bodies over time scales comparable to the age of the solar system. The analytical statistical approach of Öpik was extended to obtain a powerful Monte Carlo numerical technique by J. R. Arnold (1965). Application of Arnold's approach to a wide variety of dynamical problems in the present solar system has permitted major progress in understanding the origin and ultimate fate of planet-crossing bodies, such as Earth-approaching Apollo asteroids, comets, and meteorites (Wetherill 1985b, 1987a). In the present work Arnold's approach has been extended further not simply to include a single small "stray" body crossing the orbit of a number of planets in fixed orbits, but to investigate the simultaneous orbital evolution of a fairly large ensemble of perturbing, colliding, and growing bodies.

The fact that both computing time and memory requirements vary with the square of the number of bodies places limitations on the number of bodies in the initial state. Five hundred bodies can be conveniently handled by moderate computer systems such as a VAX Station II and about 75 simulations have been carried out for this number of bodies, employing a variety of initial states and assumptions. In a few cases, 1,000 bodies have been studied, with very similar results. Although this taxes the capacity of this computer system, there is no doubt that significantly larger ensembles ($\sim 10^4$) could be accommodated using more powerful available computers and longer computing times. Before seriously considering allocation of resources to such a "brute force" approach, I would first want to address a number of unresolved questions that could influence the extent to which the expected increase in understanding is commensurate with the investment.

Figures 2.11a-g illustrate the outcome of just one simulation. They represent "snapshots" of the evolution in mass semimajor axis and eccentricity during the course of evolution for a case that led to a final state resembling the present terrestrial planets. Even with identical starting conditions, no two simulations turn out the same. The sensitivity of the perturbed orbits to the details of a close encounter cause the system to be highly chaotic. All of the simulations have some features in common, however. These include similar final numbers of terrestrial planets (3 to 5), similar velocity and size spectra of the bodies during their growth, and the characteristic occurrence of "giant" impacts resulting from high-velocity collisions between bodies of comparable mass.

The initial distribution is shown in figure 2.11a. The largest bodies are the open circles; smaller bodies are shown as dots. Many tiny bodies are not plotted because they would overcrowd the figure. The general shape of the initial distribution may look strange.

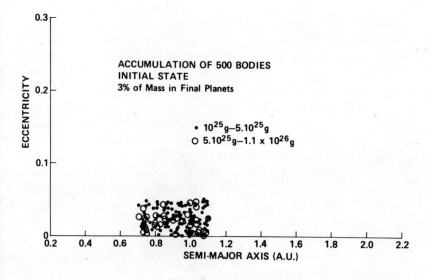

Fig. 2.11a. *Initial state. In order to avoid crowding, only half of the bodies with masses between 10^{25}g and 5×10^{25}g were plotted, and none of the 210 bodies with masses $<10^{25}$g were plotted. Solid circles represent bodies with masses of 10^{25}g to 5×10^{25}g; open circles, those with masses of 5×20^{25}g to 1.1×10^{26}g.*

The fairly sharp upper limit on eccentricity and the sharpness of upper and lower bounds of the semimajor axis are of no particular consequence. The swarm quickly evolves in such a way as to smooth these out, and the nature of the outcome is not different if they are smoothed at the outset. However, the absence of a significant number of bodies beyond about 1.1 AU and below about 0.7 AU is essential. This is not simply a characteristic of the numerical approach. This requirement can be shown to be dictated by much more general considerations of conservation of mass, energy, and angular momentum (Wetherill 1978). The final values of these quantities are determined by the present observed values, and it is an observational fact that most of the terrestrial planet mass, energy, and angular momentum is found in a narrow zone of heliocentric distance, extending not far beyond the present orbits of Venus and Earth. During the growth of the planets, mass and angular momentum are conserved, except for that associated with a very few small bodies perturbed out of the solar system at the very end of accumulation. Energy is nearly conserved, but some is lost as heat during the collisions that are essential to growth. An initial swarm that conserves angular momentum but loses energy, like a stellar accretion disk, can be shown always to spread rather than shrink in radial extent. Therefore, the initial distribution must be narrower, not wider, than the present distribution. No great harm is done if it is extended a little bit—say, from 0.6 to 1.2 AU—or if a small amount of material is placed beyond these limits. But the more this is done, the less often it is found that the final outcome resembles the present system of terrestrial planets.

How could this restricted range of semimajor axes arise? This is a good question and it deserves a good answer, but it would be premature to say much about this at present. It may not be hard to imagine that inside of 0.6 or 0.7 AU high temperatures prevented the formation of planetesimals. This does not explain the required near absence of material beyond 1.1 or 1.2 AU. This is almost certainly related to a larger problem, the gross dearth of material in the asteroid belt between the orbits of Mars and Jupiter. This region is deficient in solid material by a factor of 10^3 to 10^4, when compared with estimates of the initial distribution of matter in a plausible model of the solar nebula. There is a general supposition that nearly complete removal of this material from the asteroid belt and its considerable depletion in the region of Mars took place before the final accumulation of the terrestrial planets and is in some way associated with the influence of the gas-rich giant planets Jupiter and Saturn. A number of people, including myself, are trying to understand this set of problems at present, but I think it not unfair to say that although some good ideas have been proposed, we are still far from a satisfactory understanding of this important problem.

The state of the system after 500,000 years is shown in figure 2.11*b*. Even on this short time scale, the eccentricities of the bodies have increased to about 3 times their initial values, and the range of heliocentric distances has also become larger. The bodies have also grown; nine objects in the small-planet range (2×10^{26} to 10^{27}g) have been formed. On this figure I have also noted the positions of the objects destined to occupy the positions of the present four terrestrial planets. It can be seen that they are not exactly where they are at the present time. Again, all of the smaller bodies have not been plotted, but they fill in the region occupied by the larger bodies rather densely.

After 2.2 million years (figure 2.11*c*), two bodies of mass $>10^{27}$g have appeared. These will ultimately occupy the present positions of Earth and Venus. The eccentricities of these bodies are distinctly

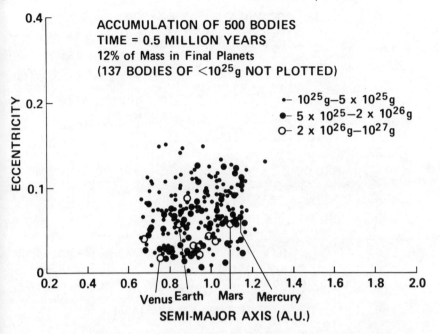

Fig. 2.11b. *Time = 0.5 million years. Small solid circles represent bodies with masses of 10^{25}g to 5×10^{25}g; large solid circles, those with masses of 5×20^{25}g to 2×10^{26}g (approximately lunar-size bodies); open circles, those with masses of 2×10^{26}g to 10^{27}g (small "planets"). The positions of the "embryos" of the final planets are indicated and named by analogy between the calculated final distribution and the observed terrestrial planets. Not plotted are 137 bodies with masses $<10^{25}$g.*

lower than the mean eccentricities of the smaller bodies. The eccentricities and inclinations (not shown) of the smaller bodies have reached a more or less Gaussian distribution corresponding to a relative velocity of about 5 km/sec, similar to the escape velocity of the largest bodies. The eccentricities of many of these bodies is large enough for their orbits to cross those of both "Venus" and "Earth." Already at this stage of growth, the concept of individual "feeding zones" for the embryos of the terrestrial planets appears to be of little usefulness. It can also be seen that fourteen bodies of mass comparable to or larger than the present mass of Mercury have been formed. The fate of many of these bodies will be collision with "Earth" and "Venus," constituting major impacts on those bodies.

After 11 million years (figure 2.11*d*), an additional body has emerged with a mass about twice the present mass of Mars. Five

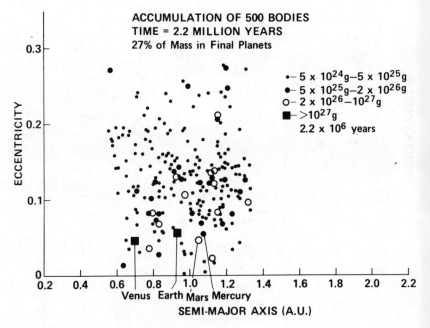

Fig. 2.11c. *Time = 2.2 million years. All bodies plotted. Small solid circles, $5 \times 10^{24}g$ to $5 \times 10^{25}g$; large solid circles, $5 \times 10^{25}g$ to $2 \times 10^{26}g$; open circles, $2 \times 10^{26}g$ to $10^{27}g$; solid squares, $>10^{27}g$ (large terrestrial planets).*

million years from that time, it struck "Venus" with an impact velocity of 13 km/sec and an energy corresponding to 65% of the gravitational binding energy of the combined system. Giant impacts of this magnitude may be expected to have quite catastrophic effects, such as ejection of material into planetocentric orbit with a mass and angular momentum comparable to that of Earth's moon. Another object of interest is "Mercury," with a semimajor axis of 1.3 AU.

At 31 million years (figure 2.11e), the number of bodies remaining is considerably reduced, almost all of them having been swept up by larger bodies. "Earth," "Venus," and "Mars" have approximately their present mass and position, whereas "Mercury" is still far out of place. During the next 3 million years, a chance series of close encounters with "Mars," "Earth," and "Venus" will cause its semimajor axes to swing all the way in to 0.58 AU. Wide excursions of bodies in this size range are not uncommon; usually they strike "Earth" or "Venus" in the process. When by chance they escape the domain of these larger planets, they survive as small planets at

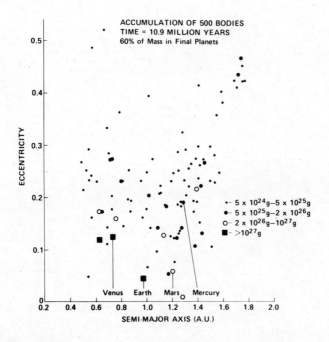

ACCUMULATION OF 500 BODIES
TIME = 10.9 MILLION YEARS
60% of Mass in Final Planets

- $-5 \times 10^{24}g-5 \times 10^{25}g$
- $-5 \times 10^{25}g-2 \times 10^{26}g$
- $O-2 \times 10^{26}g-10^{27}g$
- $\blacksquare->10^{27}g$

Fig. 2.11d. *Time = 10.9 million years. Points have same definition as in figure 2.2, curve C.*

large or small heliocentric distances. Some of the smallest bodies are in orbits extending far into the present asteroid belt. Most of these will be perturbed into solar system escape orbits. Occasionally such bodies assume more stable asteroidal orbits, and for this reason the inner solar system should be seriously considered as a potential source for some asteroidal material and the meteorites derived therefrom.

At 64 million years (figure 2.11*f*), "Mercury" has finally been perturbed into the innermost solar system. Most of the original material has now been gathered into the final planets. The last impact calculated is the collision of a $5 \times 10^{24}g$ body with "Mars" at 202 million years. The three remaining small bodies shown at 239 million years (figure 2.11*g*) were ejected from the solar system in hyperbolic escape orbits. The positions and masses of the four final planets resemble those observed in the present solar system.

Simulations resembling the present solar system as closely as those shown in figures 2.11*a-g* by no means always occur. Sometimes there are only three final bodies with mass $>10^{26}g$, sometimes there are five. The final positions vary and sometimes

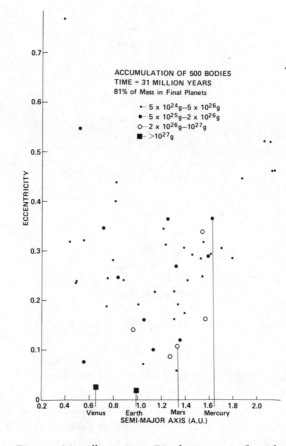

Fig. 2.11e. *Time = 31 million years. Final sweep up of residual planetesimals has begun.*

"Venus" is larger than "Earth." I suspect this variation is characteristic of actual planetary systems, and what we see in our own solar system is determined stochastically, rather than deterministically.

The phenomena of giant impacts and extensive radial migration are common features of all these simulations. These phenomena may be expected to be of more general occurrence than the range of initial states explored in this work. The mutual gravitational perturbations of the large initial bodies are insufficient to perturb any major fraction of them out of the terrestrial planet region. If the final outcome of a model of terrestrial planet formation is to con-

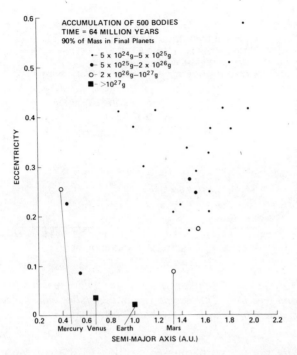

Fig. 2.11f. *Time = 64 million years. All final planets are in their final positions. A population of bodies remain, mostly with high velocities and large semimajor axes.*

tain only four final planets, and if there are more than four large bodies at the outset of the final stage of accumulation, then giant impacts will occur. The large radial excursions are a general consequence of these extra "failed planets" being accelerated to velocities comparable to the escape velocities of the final planets.

The time scale for the growth of "Earth" and "Venus" is illustrated in figure 2.12. In order to smooth out the effect of stochastic fluctuations, six simulations were averaged and scaled to the mass of these planets in the present solar system. About half the mass of the planets is accumulated in about 10^7 years. With the passage of time the growth rate decreases; about 10^8 years are required to accumulate all but the last 1% of the mass.

The occurrence of "giant impacts" the size of Mercury and Mars with Earth and Venus has already been pointed out. Figure 2.13 depicts the time and magnitude of such impacts on "Earth" for twelve simulations. During each simulation about one impactor more massive than Mars is found, as well as a similar number of

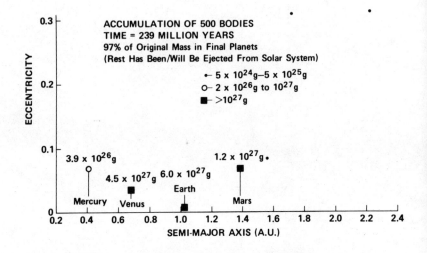

Fig. 2.11g. *Time = 239 million years. Planets have reached their final masses and positions. Three small remaining bodies will be ejected from the solar system by "Earth" and "Mars" perturbations.*

impacting bodies with masses similar to that of Mercury and a large number of bodies with masses larger than that of the moon.

CONSEQUENCES REGARDING THE INITIAL STATE OF THE EARTH AND TERRESTRIAL PLANETS

If the terrestrial planets actually formed in the way described by these calculations, some traditional (and still common) modes of thinking about the formation and early history of the Earth will require revision. For example, these results raise serious questions about the validity of the concept of individual "feeding zones" of planetesimals from which each of the terrestrial planets grew. The widespread radial migration of the planetesimals will tend to smooth out more primordial chemical differences that may have resulted from radial temperature and pressure variation in the solar nebula.

At the same time, the phenomena found in the present work suggest possible new mechanisms of chemical fractionation. It is found that giant impacts of the sort described for Earth and Venus also will occur on Mercury and Mars. The effect of such impacts on smaller planets will be quite different, however. The collision of the Earth with a body ⅕ its mass (twice the mass of Mars) may be

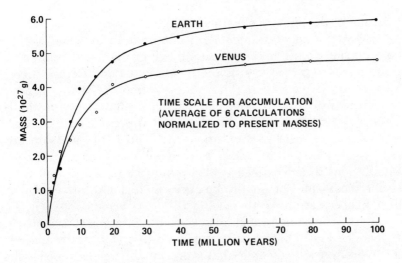

Fig. 2.12. *Time scale for growth of "Earth" and "Venus" obtained by averaging six calculations and scaling the growth to the present observed masses of these planets.*

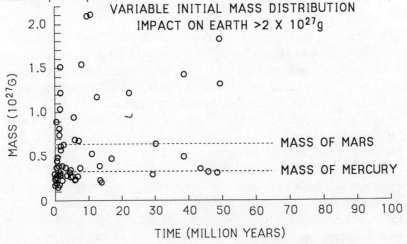

Fig. 2.13. *Giant (>2 × 10^{26}g) impacts on Earth for twelve calculations. On the average, for each case about one impact more massive than the present mass of Mars is expected, as well as a larger number of smaller but still very massive impacts.*

expected to have effects that are both catastrophic and important. Nevertheless, the Earth will survive such impacts. In contrast, when a body ⅕ the mass of Mercury strikes that planet, the combination of the higher velocities found for smaller bodies and the lower gravitational binding energy of Mercury can jeopardize the continued existence of the planet. As will be discussed below, it is quite likely that Mercury should have melted and formed a core long before it reached its final mass. If impacts of this general magnitude fail to destroy the planet, it may be that they will simply remove and shatter most of the silicate mantle, but leave the deeper iron core more nearly intact. An event of this kind could conceivably be responsible for Mercury's anomalously large iron core. Simulations directed toward identifying the occurrence of catastrophic collisions on small planets provide quantitative support for this speculation (Wetherill 1987b).

Another puzzling probable fact about Mercury's core is that it appears to be still liquid, a necessary condition for the existence of the hydromagnetic dynamo that is presumably responsible for the planet's observed magnetic field. It is hard to understand how a small planet like Mercury could avoid cooling so much during the 4.5 billion-year history of the solar system that its core would now be solid metal. This problem can at least be alleviated by assuming that Mercury's iron core is alloyed with fairly large quantities of sulfur, thereby lowering its freezing temperature. It has not usually been thought probable that Mercury should have been endowed with significant quantities of this quite volatile element if it formed as close to the sun as seemed to be indicated by its present position in the solar system. If instead Mercury accumulated material from a wide range of heliocentric distances in the manner shown in figures 2.11a-g, then one is free to propose that the sulfur content of Mercury could be similar to that of the other terrestrial planets.

If the Earth was formed by accumulation of planetesimals starting from initial conditions of the sort considered here, it appears inevitable that during almost all of its growth its internal temperature was high enough to at least partially melt its interior. Until less than a decade ago, most discussions of the thermal history of the Earth assumed that during its growth the Earth was relatively cold, unless it formed at a rate so fast as to be incompatible with accumulation of planets from planetesimals. According to this thinking, at least several hundred million years would be required for the accumulation of heat generated by radioactive decay of uranium, thorium, and potassium to elevate the Earth's temperature to the melting point and begin its geochemical differentiation and core formation. The conclusion came under question following the work of Safronov, who found that most of the energy that initially heated the Earth was not produced from radioactivity, but by the sweep up of planetesimals large enough (about 100 kilometers in diameter) to bury much of their large kinetic energy below the surface of the craters produced by their impact. Still, it was concluded that this buried energy was not likely to produce internal melting at great depth until further heating by radioactivity had been accumulated.

If much larger impacts of the size shown in figure 2.18 accompanied the formation of the Earth, it is inevitable that extensive melting should have occurred early in its growth, certainly within a few million years. If so, melting of iron and silicate, downward flow of iron, and upward flow of silicate would take place throughout the growth of the Earth and not constitute a separate, later event in Earth history. It is of interest that alternative ways of forming the Earth that are currently under consideration also lead to initial temperatures this high or higher. For example, if the formation of the Earth took place in the presence of a gas-rich solar nebula, as considered by Hayashi and his coworkers, the Earth's initial atmosphere would have been enormous, several percent of the total mass of the Earth. The thermal structure of such an atmosphere alone would lead to temperatures at the base of the atmosphere of about 2,500°K, well above the melting point of silicates and iron. If, as proposed by Cameron (1978), the Earth formed instead from a "giant gaseous protoplanet" resulting from a massive instability in a gaseous solar nebula, similar high temperatures would be expected. For these reasons geochemists and petrologists concerned with the chemical evolution of the Earth's crust and mantle should consider very seriously the consequences of a high-temperature initial state of the Earth.

Unless the early atmosphere was very massive, the average surface temperature was probably not very different from that at present. The reason is that although the internal heat flow would have been about 10^4 times its present value, the total energy input to the Earth's surface would only be about twice that at present, because the present internal heat flow is of only minor significance when compared with the energy input from sunlight. In a steady state, the surface temperature varies only with the fourth root of the energy input and the average surface effect of the larger internal heat flow would not be striking. Of course, the surficial geomorphology of the Earth would be very different than at present. Volcanoes and their gaseous emanations would be far more prolific and any thin initial crust would be continually breached by the

Earth's much more violent internal dynamics and the effects of massive impacts.

Giant impacts of the magnitude considered here were proposed more than a decade ago by Hartmann and Davis (1975) and by Cameron and Ward (1976) as a way of providing the angular momentum necessary to "launch" the moon into orbit about the Earth, an otherwise quite intractable difficulty in earlier attempts to understand the origin of the moon. Impacts of the magnitude proposed by these workers are natural and expectable events during formation of terrestrial planets in the manner described here. Giant impacts may also be expected to influence the evolution of the atmospheres of the Earth and other terrestrial planets. Cameron (1983) proposed that the approximately 100-fold excess of ^{36}Ar and neon in the atmosphere of Venus may be the consequence of the single giant impact responsible for the formation of the moon. It seems likely that the atmospheric consequences of large impacts could be even more far-reaching. The size of an impact necessary to remove most of a planetary atmosphere is not at all well-known, but it seems highly reasonable to suppose that the required size should be much less than that required to place in orbit enough of the Earth's mantle to form the moon. All the impacts shown in figure 2.13 could plausibly accomplish atmospheric removal, and possibly smaller impacts would be of importance, as well.

It is generally believed that the atmospheres of the terrestrial planets are entirely secondary, being derived from the release of gas absorbed on the primordial grains incorporated in the planetesimals from which the Earth was formed. A lingering problem with this picture has been the difficulty of quantitatively reconciling the degree to which laboratory experiments indicate inert gases can be absorbed on even carbon-rich grains and the observed concentrations of the gases in the atmospheres of the planets. This problem is compounded when one considers the extent to which outgassing should have accompanied the early terrestrial thermal history discussed above, in combination with the probability that atmospheres produced by this early outgassing were episodically removed by giant impacts.

Despite general opinion to the contrary, I have suggested that the possibility of at least some of the terrestrial planetary atmospheres being derived by gravitational capture from the residual solar nebula deserves to be considered more seriously (Wetherill 1985a, 1986). It has already been mentioned that Hayashi and his coworkers have proposed that a primordial atmosphere about 10^5 times as massive as the present atmosphere of the Earth was captured gravitationally from a solar nebula containing the full complement of hydrogen and other volatile elements that would accompany the nonvolatile matter of the Earth when mixed in cosmic proportions. A primordial atmosphere this large has not been accepted by most workers, in large part for geochemical reasons, such as the difference between the isotopic composition of neon currently being outgassed from the Earth's mantle and neon of solar origin.

It is by no means necessary, however, that consideration of gravitational capture of nebular gas be restricted to the full complement of nebular gas. In the calculations of the final stage of accumulation described in this article, the accumulation was considered to be gas-free only from a dynamical point of view. There is no doubt that more or less the full complement of gas was present at least early in the process of accumulation. If most of this—say, 90%—were removed by solar-driven outflow in about 10^6 to 10^7 years, the dynamic effects of the residual gas would be minor. Nevertheless, gravitational capture of atmospheres much more massive than the present atmospheres of Earth and Venus would be expected. It is possible that the poorly understood mechanisms by which the bulk of the solar nebula gas was removed would result in a residual nebula both chemically and isotopically fractionated.

This gravitationally captured, chemically fractionated gas would mix with outgassed volatiles from the hot interior of the planets and be episodically and stochastically lost back to the nebula following giant impacts. Stochastic differences in the extent of nebular depletion at the time of the last atmosphere-removing impact could contribute to differences between the atmospheres of the planets. Finally, late-stage addition of volatile material from bodies originating beyond the present orbit of Mars may have contributed volatile material, including distinct isotopic inert gas components, to the Earth and Venus, both directly by impacts and indirectly by mixing these volatiles with residual nebular gas followed by gravitational capture.

It will be obvious to the reader that these qualitative comments regarding the implications of the mode of terrestrial planet formation considered in this article are highly speculative and that an enormous amount of serious work is required for their evaluation. But neither should they be dismissed as far-fetched or ad hoc.

•

Earlier discussions of, for example, the chemical abundances and isotopic compositions of the inert gases in the atmospheres of Earth and Venus have been based primarily on evidence from meteoritic inert gases. Most likely, the probable asteroidal sources of these meteorites were never associated with bodies as large as these planets. Therefore, they have experienced a more limited range of fractionation processes than the bodies of planetary size that determined the nature of planetary atmospheres, including that of the Earth. In addition to those fractionation effects peculiar to small bodies formed in a different part of the solar system, fractionation processes peculiar to actual planet formation within 1.5 AU should not be ignored if a true understanding of atmospheric origin is to be achieved.

It should be obvious that it is still "early times" in the development of the appropriate conceptual basis of terrestrial planet formation, not to mention that of the remainder of the solar system. Most of the work remains to be done and the number of particular problems that can be profitably investigated is very large. But I believe it almost equally obvious that new concepts, quite likely some of those discussed in this article, will play a prominent role in the future understanding of this ancient concern of human thought—how our world began.

References

Arnold, J. R. 1965. The origin of meteorites as small bodies. II. The model. III. General considerations. *Astrophysical Journal* 141:1,536–56.

Boss, A. P. 1985. Collapse and formation of stars. *Scientific American* 252, no. 1: 40–45.

Cameron, A. G. W., and W. R. Ward. 1976. The origin of the moon. *Lunar Science* 7:120–22.

Cameron, A. G. W. 1978. Physics of the primitive solar accretion disk. *The Moon and the Planets* 18:5–40.

———. 1983. Origin of the atmospheres of the terrestrial planets. *Icarus* 56:195–201.

Chandrasekhar, S. 1942. *Principles of stellar dynamics.* Chicago: Univ. of Chicago Press.

Edgeworth, K. E. 1949. The origin and evolution of the solar system. *Monthly Notices of the Royal Astronomical Society* 109:600–09.

Greenberg, R., J. Wacker, C. R. Chapman, and W. K. Hartmann. 1978.

Planetesimals to planets: A simulation of collisional evolution. *Icarus* 35:1–26.

Hartmann, W. K., and D. R. Davis. 1975. Satellite-sized planetesimals and lunar origin. *Icarus* 24:504–15.

Hornung, P., R. Pellat, and P. Barge. 1985. Thermal velocity equilibrium in the protoplanetary cloud. *Icarus* 64:295–307.

Nakagawa, Y., C. Hayashi, and K. Nakazawa. 1983. Accumulation of planetesimals in the solar nebula. *Icarus* 54:361–76.

Öpik, E. J. 1951. Collision probabilities with the planets and the distribution of interplanetary matter. *Proceedings of the Royal Irish Academy* 54A:165–99.

Safronov, V. S. 1962. A particular solution of the coagulation equations. *Doklady Academii Nauk, USSR* 147:64–67.

―――. 1969. Evolution of the protoplanetary cloud and formation of the earth and planets. Moscow: Nauka. Translated by the Israel Program for Scientific Translations (1972). NASA TT F–677.

Stewart, G. R. and W. N. Kaula. 1980. A gravitational kinetic theory for planetesimals. *Icarus* 44:154–71.

Stewart, G. R. and G. W. Wetherill. 1988. Evolution of planetesimal velocities. *Icarus* 74:542–553.

Weidenschilling, S. J. 1980. Dust to planetesimals: Settling and coagulation in the solar nebula. *Icarus* 44:172–89.

Wetherill, G. W. 1978. Accumulation of the terrestrial planets. In *Protostars and planets*, ed. T. Gehrels. Tucson: University of Arizona Press.

―――. 1980a. Formation of terrestrial planets. *Annual Review of Astronomy and Astrophysics* 18:77–113.

―――. 1980b. Numerical calculations relevant to the accumulation of the terrestrial planets. In *The continental crust and its mineral deposits*. Ed. D. W. Strangway. Geological Association of Canada Special Paper no. 20.

―――. 1981. The formation of the Earth from planetesimals. *Scientific American* 244:162–74.

Wetherill, G. W. and L. P. Cox. 1984. The range of validity of the two-body approximation in models of terrestrial planet accumulation. I. Gravitational perturbations. *Icarus* 60:40–55.

―――. 1985. The range of validity of the two-body approximation in models of terrestrial planet accumulation. II. Gravitational cross-sections and runaway accretion. *Icarus* 63:290–303.

Wetherill, G. W. 1985a. Giant impacts during the growth of the terrestrial planets. *Science* 228:877–79.

―――. 1985b. Asteroidal source of ordinary chondrites. *Meteoritics* 20, no. 1:1–22.

―――. 1986. Accumulation of the terrestrial planets and implications concerning lunar origin. In *Origin of the Moon*. Ed. W. K. Hartmann, R. J. Philips, and G. J. Taylor.

―――. 1987a. Dynamical relationships between asteroids, meteorites, and Apollo-Amor objects. *Philosophical Transactions of the Royal Society of London,* A323–37.

―――. 1987b. Accumulation of Mercury from planetesimals. In *Mercury.* Ed. C. R. Chapman and F. Vilas. Tucson: Univ. of Arizona Press. In press.

LYNN MARGULIS

3   *The Ancient Microcosm*
of Planet Earth

Presumably about 4,500 million years ago, our
planet became a solid object and the Earth/moon system formed. In
presenting a brief overview of the early history of the Earth since
that time, I will explain some of the ways in which scientists working in this field hope they can reconstruct this early history and the
effect that the phenomenon of life wrought upon it.

There are three complementary ways of reconstructing ancient
life: through the study of stromatolites, through the study of microfossils, and by means of organic geochemistry. Stromatolites are
laminated rocks known for many years to geologists as domed, layered limestones. They are like coral reefs in that, just as there can be
no coral reefs without previous communities of coral animals, there
are no stromatolites without preexistent communities of bacteria.
Stromatolites, like the reefs, are fossils of entire communities of
interacting organisms.

A second way we reconstruct history is by the direct study of
microscopic fossils. Fossils of the microcosm require microscopic

This chapter was presented as a talk at the April 1986 meeting of the
National Academy of Sciences. I thank Dorion Sagan, who converted this
manuscript from the spoken to the written word, and John Kearney, who
rendered it increasingly legible. We acknowledge with gratitude research support from the NASA Life Sciences office (NGR–004–025) and the Lounsbery
Foundation.

techniques for their visualization. The best preserved microfossils tend to be found in smooth, black, siliceous rocks called cherts. Cherts (figure 3.1), cryptocrystalline quartz, are extremely important for reconstructing the early history of life. An aspect of their genesis is related to our studies of modern microbial mats.

A third method utilizes organic geochemistry. Because there is much ambiguity in the interpretation of the organic geochemical evidence of early life, I will avoid this area in my discussion. However, organic geochemistry shows much promise; extraction of organic matter in rocks yields compounds that are remains of the earliest interacting microorganisms and sedimentary materials (Hayes, Kaplan, and Wedeking 1983).

We consider stromatolites and microfossils to be direct remnants of life in the early record. Though these are always associated with sedimentary rocks, we must always date them, using geochronological techniques, by the igneous rocks that intrude upon them. Even then the dates are always only approximations. But from these dates we can obtain minimal ages for the rocks of biological interest.

Fortunately, the fossil history of life is to some degree reflected in the present. We can turn to extant environments as a source of information for the interpretation of early life. There are not many ecological settings on today's Earth that entirely lack the latecomers we call plants and animals but are host to many varieties of microorganisms. There are a few, however, and they are accessible to investigators. Modern bacterial communities serve us as models for understanding how stromatolites formed and how some original bacterial remains were trapped and preserved in cherts.

It is important to realize that the whole field of study of the ancient microcosm of planet Earth is relatively new. The microfossil studies began with the late Elso Barghoorn, professor of botany at Harvard University until his death in 1984. Barghoorn transcended the limits of his field and discovered a "flora" so old that it antedated plants and animals. All of us in this new area of microcosmic research owe a great debt to Barghoorn, who inaugurated Precambrian paleobiology. Barghoorn began working in collaboration with Professor Stanley Tyler of the University of Wisconsin in the early fifties (Barghoorn and Tyler 1965). It is obvious that we owe a debt to subsequent workers, too, many of whose contributions appear in Schopf (1983).

Most scientists interested in the question agree that life on Earth

Fig. 3.1. *Sample of black chert from the Kromberg Formation in southern Africa, part of the Swaziland System, over three billion years old.*

began more than 3,500 million years ago, and perhaps long before that. This assertion is based on the fact that from 3,500 million years ago until the present, we have a continuous fossil record of life. Although the early part of this record consists primarily of trace fossils (such as stromatolites) and of microfossils, it nonetheless exists. Creationists are simply misinformed when they maintain that there is no record of early life. It has waited only to be recognized. Many talented graduate students are uncovering more of it every year.

The study of early life is primarily that of prokaryotic, or bacterial, cells, which are metabolically complex but structurally simple cells. These cells differ from all other kinds of cells—that is, from the cells of animals, plants, fungi, and protoctists—in that they lack nuclei.

Some 580 million years ago, the first animals with hard parts appeared. It is in that time, the beginning of the Phanerozoic eon, that the much better-known fossil record begins. The Phanerozoic eon includes the Paleozoic, Mesozoic, and Cenozoic eras. Visitors to the halls of ancient life in most natural history museums are exposed essentially only to this last fraction of Earth history. Until recently, the pre-Phanerozoic (that is, the Precambrian, comprised of the Archean and Proterozoic eons) has been mute.

The question is: what happened during the entire stretch of time that preceded the Phanerozoic eon? Nearly everyone who has examined this problem agrees that the early portion of this great stretch of time, the Archean eon, should be called "the age of bacteria." We can probably add that the Archean, from 3,500 until 2,500 million years ago, was the age of anaerobic bacteria; that is, it was most likely dominated by bacteria that did not directly use oxygen in their metabolism. The earliest form of photosynthesis was not oxygen-producing. Later, oxygenic, photosynthetic bacteria appeared and released oxygen into the atmosphere. During this later part of the age of anaerobic bacteria, many complex communities of bacterial cells emerged. Bacteria mutated, exchanged their viruses and plasmids, attacked each other, and sometimes merged with one another. In general, bacteria interacted and diversified, eventually producing the larger, more familiar organisms composed of nucleated cells that include those of plants and animals (Margulis and Sagan 1986a).

These organisms composed of nucleated cells—protoctists, fungi, animals, and plants—came relatively late to life on Earth (Margulis and Sagan 1986b). Most of the earlier life was that of bacteria living in well-structured communities. Life on Earth today is still dominated by such bacteria.

The laminated rocks that geologists recognize as stromatolites represent the remains of just such actively growing communities of bacteria. Among the earliest remains of bacterial communities are the Warrawoona stromatolites from western Australia (figure 3.2). Formed 3,400 million years ago, they are, with the South African trace fossils, among the oldest stromatolites on record. Stromatolites can reach very large sizes. Figure 3.3 shows one from northwest Canada that is about 2,300 million years old. We now know

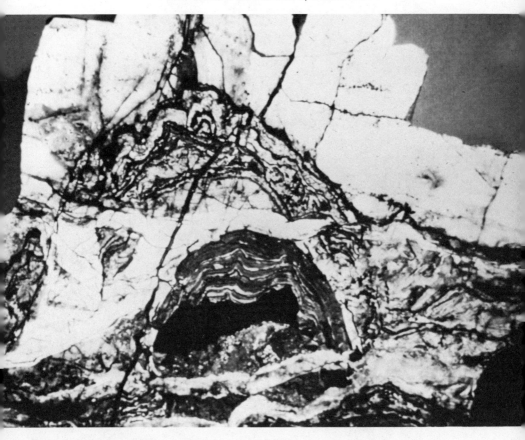

Fig. 3.2. *Warrawoona stromatolite from western Australia.*

that structures of this kind were made in intertidal, subtidal, or supertidal environments, presumably by sustained communities of bacteria over long periods of time. Most stromatolites are domed, layered, calcium carbonate rocks, though sometimes they are made of chert. Calcium carbonate stromatolites are remains of bacterial metabolic processes: they derive from the trapping, binding, and growth activities of communities of organisms. The major builders of these kinds of laminated limestone rocks were cyanobacteria, the oxygen-producing, phototrophic bacteria once called blue-green algae.

Other stromatolites are known from the Great Slave Lake area in the Northwest Territories of Canada, a few hundred miles below

Fig. 3.3. *Proterozoic stromatolite from Victoria Island, Northwest Territories, Canada. Note figure of man at lower right.*

88

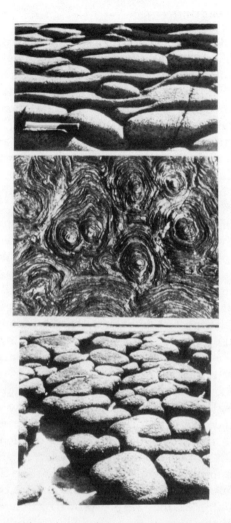

Fig. 3.4. *Ancient and modern carbonate stromatolites compared.* Top: *Proterozoic carbonate stromatolites from the Great Slave Lake, Northwest Territories, Canada. The geology hammer is about one foot long. A Pleistocene glaciation groove is at the lower right.* Middle: *These two-billion-year-old stromatolite heads from the Great Slave Lake were cut by Pleistocene glaciation in this bedding plane exposure. The internal laminated structure of these pre-Phanerozoic stromatolites is very similar to that of recent ones.* Bottom: *Recent carbonate stromatolites in a warm, hypersaline environment: Hamelin Pool, Shark Bay, western Australia. Sediment particles are trapped and bound by the microbes, primarily oxygenic photosynthetic prokaryotes with well-developed sheaths. They are cemented by the precipitation of inorganic aragonite. These modern lithified intertidal stromatolites contain thriving communities of microorganisms, including ones that bore into carbonate and decrease the rate of growth of the heads.*

the Arctic Circle (figure 3.4). We know that they were laid down between 2,000 and 3,000 million years ago in environments that were then tropical and subtropical in climate.

At least one extant setting, roughly comparable to the kind of environment in which fossil stromatolites developed, is found in western Australia at a place called Hamelin Pool, in Shark Bay. These living stromatolites, shown in figure 3.5, have been under construction and growing to their present size over the last two centuries. Even today they abound in live microorganisms. The large and conspicuous Shark Bay stromatolites provide analogs of Proterozoic ancient ones.

The bacterial fabric from which the stromatolites were formed was originally soft. By extrapolation from recent work on modern communities of stromatolites in the process of forming, we can conclude that stromatolites in the fossil record also began as soft-textured microbial mat communities. Bacterial communities today, with their abundance of mucopolysaccharides, proteins, nucleic acids, chlorophylls, and other organic materials, trap and bind sand and evaporite clasts and other fine minerals, thus preserving sediment. Eventually they lithify. So, we may assume, did Archean communities.

Many microbial mat communities today do not persist at all, for many ecological or geological reasons. Some are eaten back by animals and microbes nearly as rapidly as they are produced. Some are destroyed because we build beach houses on them, for example, on the Cape Cod shore of Massachusetts. A narrow ribbon of microbial mat communities extends from Halifax, Nova Scotia, south to North Carolina along the east coast of America. Unfortunately, most are outcompeted: they are overgrown by that familiar "mammalian weed," the human race. We have severely disrupted the continuous microbial mat cover that otherwise would run down the entire coast. This tattered community, looking superficially like patches of dirty sand, nonetheless remains. Under conditions favorable to growth, its presumed ancestors thrived in the pre-Phanerozoic eon. These ancient communities spread over vast areas prior to the Phanerozoic eon when mammals and other impediments to their growth had not yet evolved. Mat communities of various forms and sizes became very extensive, especially in the middle and late Proterozoic eon, from 1,500 to 580 million years ago.

What organisms construct microbial mats that may become stromatolites? A photograph of a stromatolite-building cyanobac-

Fig. 3.5. *Shark Bay, western Australia: recent stromatolites.*

terium, or "blue-green algae," is shown in figure 3.6. (We know now that it is really neither algal nor necessarily aquamarine in color.) It is only one of thousands of kinds of oxygen-producing cyanobacteria. Only 4 to 5 microns across (many are smaller), some cyanobacteria, like that in figure 3.7 (*Microcoleus cthonoplastes*), have very extensive sheaths which tend to trap, bind, and collect sediment into rocklike structures. Their communities can be thought of as potential stromatolites.

Fundamentally, the cosmopolitan mat builders (figure 3.7) form a set of filaments like a telephone cable; such cyanobacteria are comparably sheathed. Most important, these small, long, penetrating threads of colored organisms produce oxygen and, in many cases, simultaneously precipitate calcium carbonate. They gener-

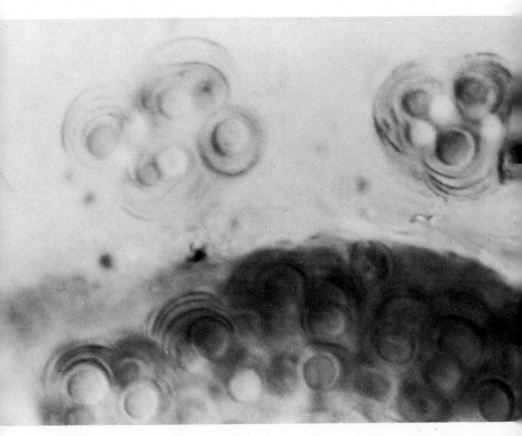

Fig. 3.6. *Coccoid mat-building cyanobacterium:* Entophysalis *sp.*

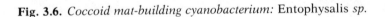

ate limestone by the removal of carbon dioxide from the atmosphere.

In some cases, chert is found in association with the remains of these organisms, either primarily—that is, originally in the environment of deposition—or perhaps secondarily to the calcium carbonate. An outcrop of chert is shown in figure 3.8. Chert, a kind of cryptocrystalline quartz of which "gunflint" is an example, is chemically silicon dioxide. Luckily, it is a superb embedding or preserving medium for microorganisms under certain conditions. Although paleobiologists may cut through five hundred black, seemingly promising samples of it and find nothing, extraordinarily well-preserved microorganisms can sometimes be found on the five-hundred-and-first sample. Most of the microfossil discov-

Fig. 3.7. Microcoleus cthonoplastes.

eries have been made by thin-sectioning chert. Almost none come
from the carbonate stromatolites, which are too porous to preserve
microorganisms well. Although the limestone stromatolites extend
very far back in time and are conspicuous reminders of the geological importance of ancient bacterial communities, they usually preserve these communities very poorly.

The cherts, on the other hand, are, on an absolute scale, an
excellent medium for preservation. Figure 3.1 showed a sample of
black chert from the Kromberg Formation of southern Africa. Sur-
prisingly to those who examine them, the amorphous, smooth
black regions in the cherts, seemingly just pure rock, tend to yield
the best microfossils. One can become extremely frustrated cutting
through many, many rocks and finding nothing of interest. For-

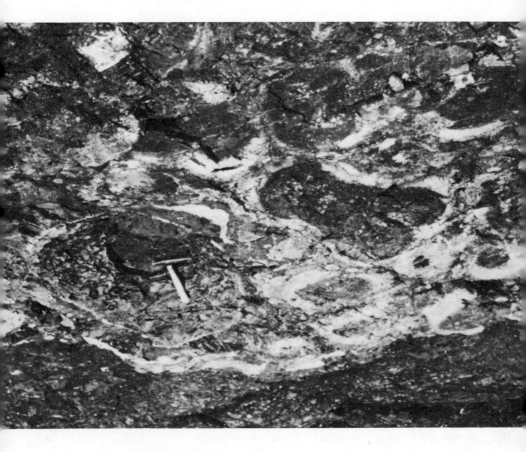

Fig. 3.8. *Gunflint chert outcrop.*

tunately, hammer scars can markedly increase the chances of discovering good microfossils in chert: the scars mark the successes of earlier explorer-geologists. Following in the footsteps of one of the masters who has hacked into the best rock samples, one can often find well-preserved ancient remains of bacteria.

The earliest preserved microfossils in the fossil record are dated from about 3,500 to 3,100 million years. Both in South Africa (in the Swaziland System of sedimentary rocks, including the Kromberg Formation, the Fig Tree Formation and Swartkoppie Zone) and in western Australia in the Pilbara Block, fossils have been found in cherts, though the quality of preservation is not spectacular. Thus, we now have both stromatolites and microfossils from two very different places on Earth that reveal records of ancient life. Figure

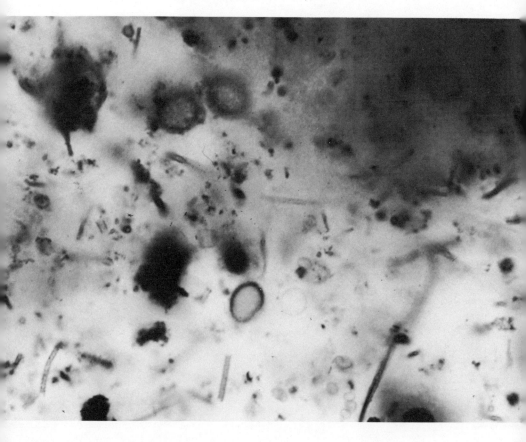

Fig. 3.9. *Microfossil from the Gunflint chert, light micrograph (×320).*

3.9, a photomicrograph taken with an ordinary light microscope under high power, shows a rock sample that has been sectioned for observation. Many histologists and cell biologists are unable to make cell preparations on microscope slides as beautiful as these naturally preserved cell materials.

More recently deposited microfossils are even better preserved. Some fossils nearly a billion years old are seen in figure 3.10. Their condition is so good one can say that, in general, these are remains of ancient cyanobacteria, the kinds of oxygenic, photosynthetic microorganisms that are involved in the production of lithified structures today. They are, in short, fossil remains of the inhabitants, perhaps even the creators, of stromatolites.

In figure 3.11a and b, we see a banded iron formation as an

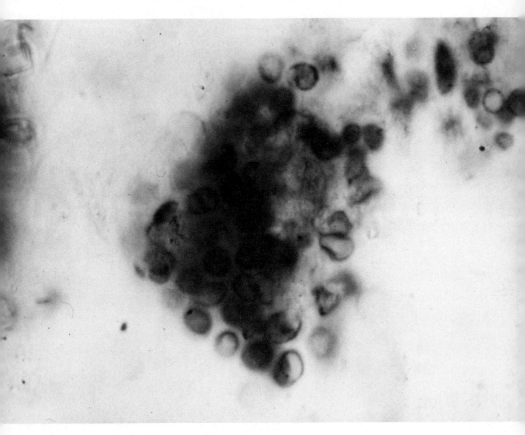

Fig. 3.10. *Chert from the Narssarssuk Formation, northwest Greenland, 700 to 800 million years old (×320).*

outcrop. These rocks have been severely metamorphosed. Had live or fossil microorganisms been associated with the sediments that became these rocks, they could not be detected because the rocks have since been heated to such high temperatures. Yet sedimentary and much less metamorphosed samples from the same locality gave birth to the burgeoning field of paleobiology. In the early 1950s, Stanley Tyler telephoned Elso Barghoorn and invited him to identify some microscopic objects in rock samples associated with a banded iron formation. In some places, for example, in this great iron deposit known as the Gunflint Iron Formation, on the northern shores of Lake Superior, Tyler had found areas where exposed cherts never subjected to metamorphosis lend themselves to study.

Fig. 3.11. *Banded iron formation:* a, *Mt. Stevens, of the Hamersley Range, western Australia;* b, *close-up of outcrop from Swartkoppe Zone of the Fig Tree Formation of southern Africa.*

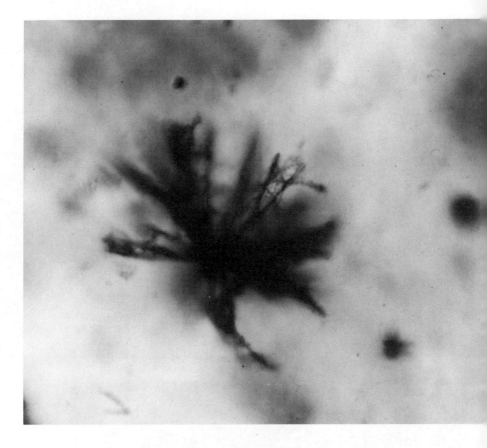

Fig. 3.12. Eoastrion *(Kakabekia): microfossils from a petrographic thin section from the Gunflint chert (×320).*

Ninety percent of the world's minable iron comes from formations like these, most of which are roughly 2,000 million years old. These great formations, even if subsequently metamorphosed, were originally sedimentary rocks. They are of great economic interest— and equally interesting in their possible origins.

The procedure for studying microfossils is relatively straightforward: one takes a piece of chert, cuts a thin sample with a diamond saw, and mounts it on a microscope slide. What is first noticed is that the banding pattern persists to below the centimeter level. When we employ higher magnifications, we see the same pattern— dark bands alternating with light—at an even smaller dimension.

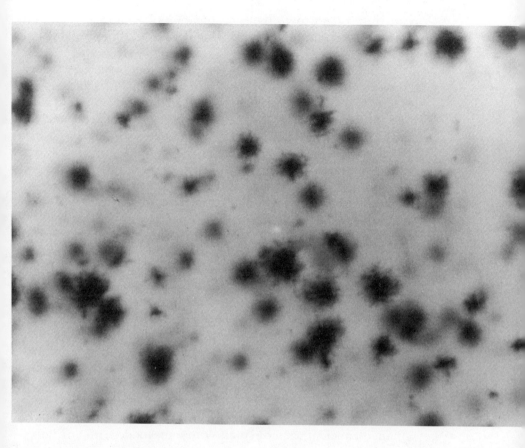

Fig. 3.13. *Modern bacterial colony of* Metallogenium, *showing resemblance to* Kakabekia.

In this sample there are about a million microfossils per cubic centimeter of rock. Barghoorn named them *Eoastrion,* meaning "dawn star." According to Paul Strother, *Eoastrion* is a form of *Kakabekia;* thus, *Kakabekia* is the preferred name (figure 3.12). At still higher power, they no longer seem merely rather spherical, but appear much more stellate, with pointed extensions. No one claims to be able to definitely identify these microfossils with any living organisms, but they are interpreted to be bacteria that precipitate metal. Manganese dioxide and iron oxides are often precipitated today by bacteria that grow in colonies, such as those shown in figure 3.13. Precipitation by communities of bacteria is probably related in some way to the concentration of such metal oxides in sedimentary environments over long periods of time (Holland and Schidlowski 1982).

Throughout the early history of Earth there existed an enormous quantity of life, widespread but nearly all at the microscopic level. In the Gunflint chert there are 2,000-million-year-old rocks from which we obtain direct and detailed evidence of ancient, dense populations of bacteria. Some of the structures seen in a petrographic thin section of one of the samples are spheres and filaments that look very much like iron-precipitating bacteria taken from modern iron springs. Thus, there is a direct fossil record of oxygen-producing and iron-precipitating microbes.

Black mud, to us, is very interesting. We are now testing the hypothesis that certain gelatinous, sulfurous muds are precursors of the type of black chert in which microfossils are sometimes preserved (Stolz and Margulis 1984; Stolz, Margulis, and Guardans 1987). Since 1977 we have been working on microbial mats—colorful muds—from Cuba, Mexico, and Alicante, in Spain. Figure 3.14a, a cross-section of the top two centimeters of the microbial mat (a potential stromatolite), shows laminations. Figure 3.14b is a photograph of a sedimentary rock, interpreted to be a fossil microbial mat, which comes from 3,400-million-year-old outcrops in South Africa. The similar-looking sample in figure 3.14a is a modern microbial mat, a community of highly organized, mud-dwelling, sediment-binding bacteria. The figure compares rocks thought to be remains of ancient bacterial mats with some modern mats in northwest Mexico.

Near the Pacific coast of Baja California at San Quintin, Mexico, is the North Pond of Laguna Figueroa, some 250 kilometers south of San Diego. East of the salt marsh vegetation and the dunes toward the now quiet volcanos lies an expanse of evaporite flat that includes microbial mats. The dunes act as a barrier to the open ocean. (Elsewhere other sorts of barriers have the same effect: at Matanzas, in Cuba, the barrier is a mangrove stand.) There are many types of mats here, but we have concentrated on a kind that is dominated by the species of cyanobacterium known as *Microcoleus cthonoplastes*. These mats always form near the dunes where there is tidal channeling (figure 3.15). Apparently, this type of mat must be exposed to the quiet onlapping and offlapping of a tidal cycle.

When *Microcoleus*-dominated microbial muds dry, they produce structures called desiccation polygons. A section through such a polygon reveals that the top layer is composed of evaporite min-

a

b

Fig. 3.14. *Modern and ancient microbial mats:* a, *North Pond, Laguna Figueroa, Baja California Norte, Mexico;* b, *an ancient laminated chert from the Fig Tree Formation of southern Africa.*

101

Fig. 3.15. *A modern microbial mat, North Pond, Laguna Figueroa, Baja California Norte, Mexico.*

erals. Underneath this layer is an active mat-building community of cyanobacteria. The cyanobacteria are responsible for the cohesiveness of the mat. These cyanobacteria have sheaths, though they often lose them in laboratory culture. As they glide and grow, they take up bits of carbonate, sand, and other particles. Sectioning through their filaments reveals thylakoid membranes. These are the cell membranes of photosynthesis. At these membranous sites carbon dioxide is removed from the air while oxygen is simultaneously released. With the *Microcoleus* are other microbes: *Spirulina*, *Oscillatoria*, and some nitrogen-fixing cyanobacteria. All are capable of removing atmospheric carbon dioxide by converting it into organic or carbonate carbon. They produce oxygen, as well. Beneath these cyanobacteria are purple, phototrophic bacteria. We are still in the process of identifying many of these complex communities. The mats are so dense that they are analogous to living tissue of animals. We must dissect, dilute, and open up the structure with delicacy to study it.

Eukaryotes, organisms composed of cells with nuclei, are only minor components of mat communities. Mats are bacterial—that is, prokaryotic—structures. However, some diatoms, nucleated organisms with siliceous coverings, do inhabit the Baja California mat, especially on the surface. Almost no animals live in association with bacterial mats, but those that do are a curious lot. One exception is a nematode, a tiny worm which sometimes turns purple because it has eaten purple phototrophic bacteria. The nucleated microbes that inhabit the mat do not really contribute to its building. One member of the community is the sulfide-requiring ciliate *Metopus contortus*. Covered with bacteria on its surface, it has highly modified mitochondria—or it may not have them at all (Dyer 1984). We also find the enigmatic *Paratetramitus jugosus*, an amoeba. This bacteriovore can grow to cover a petri plate in two days. Identified as an amoebomastigote, it very infrequently seems to divide by mitosis. Instead it may indulge in some sort of growth by multiple fission. *Paratetramitus*, which has several different stages in its life cycle, will eat through colonies of bacteria with iron-coated spores. All this activity takes place in the top two centimeters of the mat.

In 1979 and 1980 enormous quantities of spring rain flooded the mat for months. The mat was lost to sight; we thought it was destroyed. Finally we found dormant mat under several meters of water. The mat community was undergoing dramatic changes. *Microcoleus*, the organism that had previously dominated the entire community, disappeared. The inundated mats retained their laminations, but a very different set of organisms now comprised them. The dominant organisms during the flood were purple phototrophic bacteria that oxidized hydrogen sulfide to sulfur or sulfate in the sulfide-rich muds. There was an influx of terrigenous sediment, including volcaniclastics from neighboring volcanoes and sand dunes. My former student Dr. John F. Stolz (now at the Department of Biochemistry at the University of Massachusetts, Amherst) found many new kinds of bacteria which had not yet been described (Stolz 1984). There was an enhanced precipitation of manganese dioxide and iron oxides by bacterial colonies. For two years the water was as fresh as that of freshwater ponds and lakes. However, during the third year, 1981, the water drained and evaporated as conditions returned to normal. Periodic exposure to tidewater and desiccation was reestablished. At first we took the two-year-long, Atlantis-like event to be tragedy, but it turned out to be a blessing in disguise. The massive floods added much to our understanding of microbial preservation. During the flood, as we contemplated the vast expanse of water under which the small mats were struggling, I had feared that we would never find the buried *Microcoleus* mats that we had been studying for some years. Our friend the sedimentologist Claude Monty had proffered his wisdom: "If you believe these mats to have been preserved for the last 3,000 million years in the fossil record, what can a little spring flood do to them?" So in the spring of 1980 we had constructed a crude boat and paddled about in the 3-meter-deep water for four hours until at last we found the laminated mats entirely submerged.

The 10 centimeters of laminated sediment shown in figure 3.16 probably represents about a century of accumulation by mat growth before the flood. In 1981 the weather conditions reverted to what they had been for the hundred or so years before the period of the storm (1979–80). The area, in the spring of 1981 and again in 1982, was like a giant petri plate; with the sun shining, bacterial colonies appeared everywhere underfoot. We feel confident that eventually the entire stratified community of *Microcoleus* will reestablish itself. It has taken nearly six years to recover completely the preflood tidal pattern and climatic conditions. We expect the entire cycle to require a decade: from desiccated, well-developed polygonal communities dominated by *Microcoleus* through freshwater flood and sedimentation back to well-developed polygons.

One of the processes we are looking for in mats is known to occur in dense microbial communities (Guerrero et al. 1986). This is the predation of one bacterium by another type of bacterium. Some bacteria actually attack others, then penetrate and divide inside of their cytoplasm. This type of interaction supports our notion of the symbiotic origin of certain parts of animal and plant cells: there is now good evidence that organelles such as respiratory mitochondria and photosynthetic plastids began as bacteria inside other cells (Margulis 1982). Multicellular, spore-forming, and other complex bacteria are also found in the mats.

In studying the fossil material, pioneer researchers have tended to concentrate on those hand samples of chert that are smooth and black. Paleobiologists know by experience that the preservation of microfossils has higher probability in these smooth black regions. Accordingly, we have treated fresh, unsilicified gelatinous black mud as if it were a fossil chert. John Stolz has found filaments and coccoids preserved in the anaerobic, very fine-grained, smooth, sulfur-rich muds. Some of them bore striking resemblance to the

Fig. 3.16. *Mat submerged by flooding.*

ancient fossils. We now feel that if the mud created during the years of storm shortly afterward underwent chertification (lithification), we might understand how, under favorable conditions, some microfossils have remained intact for some 3,000 million years after being buried. Our current work, the subject of a Ph.D. thesis by Michael Enzien, focuses on the stages from black mud to black chert: how does the sulfurous mud become silicified and therefore preserved? We do not know.

Microbial mats past and present represent one tiny branch on that mysterious tree whose trunk, studied primarily by organic chemists, is the origin of life. We know even less about that "event"—the origin of life—than we do about chertification of microbial mats. We generally assume that it is equivalent to the origin of the first bacteria and that the process occurred on the early Earth (Fleischaker and Margulis 1986). By about 3,500 million years ago, bacterial communities already seem to have been well diversified. We now have a microbial fossil record that begins with the earliest sedimentary rocks that remain undamaged. All that have not been badly metamorphosed, not badly treated by high temperatures and pressures, contain evidence of life. As early as the record yields rocks suitable for study, it yields a direct record of life in those rocks. Naturally that record becomes much clearer as we near the present.

We have learned that for the first 2,000 or possibly 3,000 million years, microbial life was thriving on the Earth. The prokaryotes, that great group of organisms known as "the bacteria," produced and cycled many of the reactive gases of the atmosphere. They also laid down sediment as limestone and were at least partly responsible for generating certain metalliferous deposits. Beginning at about 1,000 million years ago (plus or minus 300 or so million years), we detect the origin of a different kind of bacterial community organization that later becomes recognizable as a new kind of cell. At this time the Earth's atmosphere had become quite rich in free oxygen, which it had begun accumulating some 2,000 million years earlier as the result of cyanobacterial activities. The production of bacterial oxygen accelerated the oxidation and deposition of certain metals. The large-scale removal of atmospheric carbon dioxide by phototrophic and even heterotrophic bacteria led to the precipitation of even larger amounts of limestone. These sedimentary and atmospheric processes were clearly related to the diverse metabolic activities of large numbers of microbes. The forms of bacterial life that performed these feats still grow, interact with sediment, and exchange gases today. Although some bacteria have evolved into large eukaryotic forms—animals and plants—many have maintained their original prokaryotic or bacterial design (Sonea and Panisset 1983). In spite of our compulsive anthropocentrism, the ancient microcosm, whether or not we ignore it, is still a tremendously potent force (Margulis and Sagan 1986b). The microcosm is, geologically speaking, profoundly active on a planetary scale. The changing microcosm glimpsed from its indelible record in smooth black chert and stromatolitic limestone continues to maintain the cosmically anomalous chemistry of the Earth's crust, atmosphere, and oceans.

References

Barghoorn, E. S., and S. Tyler. 1965. Microorganisms from the Gunflint chert. *Science* 14:563–77.

Dyer, B. D. 1984. Protoctists from the microbial communities of Baja California, Mexico. Ph.D. thesis, Boston University.

Fleischaker, G., and L. Margulis. 1986. Autopoiesis and the origin of bacteria. *Advances in space research* 6, no. 11:53–55.

Guerrero, R., C. Pedros-Alio, I. Esteve, J. Mas, D. Chase, and L. Margulis. 1986. Predatory prokaryotes: Predation and primary consumption evolved in bacteria. *Proceedings of the National Academy of Sciences* 83:2,138–42.

Hayes, J., I. R. Kaplan, and K. W. Wedeking. 1983. Precambrian organic geochemistry, preservation of the record. In *Earth's earliest biosphere*. Ed. J. W. Schopf. Princeton: Princeton Univ. Press.

Holland, H. D., and M. Schidlowski. 1982. *Mineral deposits and evolution of the biosphere*. Dahlem Konferenzen, Berlin, West Germany: Springer-Verlag.

Margulis, L. 1982. *Symbiosis in cell evolution*. New York: W. H. Freeman.

Margulis, L., and D. Sagan. 1986a. *Origins of sex*. New Haven: Yale University Press.

———. 1986b. *Microcosmos*. New York: Summit Books.

Schopf, J. W., ed. 1983. *Earth's earliest biosphere*. Princeton: Princeton Univ. Press.

Sonea, S., and M. Panisset. 1983. *A new bacteriology*. Boston: Jones and Bartlett.

Stolz, J. F. 1984. Fine structure of the stratified community at Laguna Figueroa, Baja California, Mexico. II. Transmission electron microscopy as a diagnostic tool in studying microbial communities *in situ*. In *Microbial mats: Stromatolites*. Ed. Y. Cohen, R. Castenholz, and H. O. Halverson. New York: Alan R. Liss.

Stolz, J. F., and L. Margulis. 1984. The stratified microbial community at Laguna Figueroa, Baja California, Mexico: A possible model for pre-Phanerozoic laminated communities preserved in cherts. *Origins of Life* 14:671–79.

Stolz, J. F., L. Margulis, and R. Guardans. 1987. La comunidad microbiana estratificada de la Laguna Figueroa, Baja California, México: Un posible modelo de comunidades laminadas y microfósiles pre-fanerozoicos preservados en pedernales. *Studia Geologica Salmanticensia* 24:7–24.

was that the iridium anomalies were produced by a large-body impact, either a comet or asteroid. More important, the Alvarez group went on to claim that the coincidence in timing between the mass extinction and the iridium anomalies shows that the comet or asteroid impact caused the extinctions.

Since 1980 a number of other, supplementary hypotheses have been suggested, some of which imply yet further cosmic connections. The most celebrated of these is the hypothesis that our sun has a dim companion star, called Nemesis, which is responsible not only for the terminal Cretaceous event but also for other extinction events in Earth history (Whitmire and Jackson 1984; Davis et al. 1984).

The original Alvarez paper, supplemented by the other interpretations, has created a tremendous amount of activity. At least 350 related papers have been published since 1980 and controversy is rampant. Many people are angry, while others are enthusiastic about the new proposals. To some, the new work is symptomatic of a science gone mad. Others believe that the fields of geology, paleontology, geochemistry, and evolutionary biology are finally realizing that the Earth is not alone—that there have been a lot of things whirling around above us for a long time and that many of these may have had significant effects on the physical and biological history of the Earth. Before getting into these areas in a bit more detail, I will make some general observations about the evolutionary role of extinction. I have already noted that the extinction of species has been almost as common as origination. It follows, therefore, that extinction must play a major role in the evolutionary process. The question is: what is its role?

ROLE OF EXTINCTION IN EVOLUTION

The traditional view is a rather Darwinian one: that extinction serves as a selection mechanism at an organizational level higher than the local breeding population. The less well-adapted species or groups of species are eliminated, perhaps through active competition from the better adapted organisms. A common view of the extinction of dinosaurs is that the mammals of the late Cretaceous were more intelligent and therefore deserved to "inherit" the Earth. Will Cuppy, the famous satirist of the thirties and forties, described this by saying: "The Age of Reptiles ended because it had gone on long enough and it was all a mistake in the first place" (Cuppy 1983, 93). As so often happens with satire, Cuppy captured the conventional wisdom: extinction was seen as a good thing. This explanation of extinction may well be true, but there is increasingly little documentary support for it. In nearly all cases, perhaps in every case, the only evidence for the inferiority of an extinct organism is the fact of its extinction.

This leads us to look for other possibilities. For example, extinction may be a basically random process. Species may live in a "field of bullets" such that their extinction is purely a matter of chance, bearing no relation to their adaptive success. Leigh Van Valen proposed this about 15 years ago with his Red Queen hypothesis; he presented a considerable amount of statistical data from the fossil record to support the idea that species go extinct on a purely chance basis (Van Valen 1973). If this were true, one could legitimately describe durations of species in terms of a half-life, by analogy to radioactive decay. There is a problem with a field-of-bullets-way of looking at extinction, however. It has become clear, as better statistical analysis has been done on some of the larger events, that the extinctions are selective rather than random. Many groups of organisms had extinction rates far higher or lower than would be predicted from the purely random model.

If the extinctions have been selective, this still does not force us back to the Darwinian style of survival based on fitness. We may be dealing with what has been called "nonconstructive selection" (Raup 1986). The mass extinctions may have been caused by environmental stresses so bizarre and so rare that they were not within the experience of the organisms and thus did not relate in any way to the adaptive value or fitness of the organisms.

Let me illustrate this with a totally hypothetical example. Suppose our present terrestrial biosphere were hit by a heavy dose of high-energy radiation. We know enough about the somatic effects of such radiation to be able to select a dosage that would kill off all exposed mammals but leave most insects and plants unscathed. If the terrestrial biota were subjected to this environmental shock, we would have a highly selective extinction—selective in that some whole groups would survive and others would not—but the lists of victims and survivors would make no sense in terms of the adaptive problems normally faced by the organisms. The insects might inherit the Earth, but not because they were superior in competition with mammals.

It is possible to look at the dinosaur extinction in the same way. The large reptiles may not have done anything wrong but simply have had the bad luck to be in the wrong place at the wrong time—

perhaps susceptible to a sort of rare environmental shock which the Darwinian system of adaptation can not recognize or cope with because it is not a consistent part of the environment.

The bottom line is that we really do not know yet what the causes and consequences of extinction have been in the total evolutionary process. It is increasingly clear, however, that extinction has had large effects. Most of the mass extinctions have been followed by intervals of low diversity in which previously dominant groups of organisms are totally absent or reduced to a few species. And the rediversifications that follow include some of the most significant innovations in the history of life. It is as if global biology were being "recharged" by mass extinction. Without this effect, life on Earth might have developed very differently: it might have reached a steady-state condition early and never achieved what we now know as advanced life.

MASS EXTINCTION AND COSMIC CAUSES

Let me return to some of the details of the late Cretaceous event and to other mass extinctions. The Alvarez group used iridium anomalies in Europe and New Zealand as evidence for a collision between the Earth and a large asteroid or comet. The iridium evidence alone would have convinced most people that such an event occurred. We know that the Earth has been and continues to be bombarded by extraterrestrial bodies. There are records of about a hundred craters, some of which are very old and very large. The data from these craters and from the existing population of asteroids in Earth-crossing orbit suggest that there have been perhaps twelve collisions with bodies 10 kilometers in diameter or larger and many more collisions with smaller bodies during the past 600 million years (Shoemaker 1984). So, to find evidence of another one merely adds to an already long list of impact events. But the suggestion that one of them was the cause of a mass extinction has produced considerable furor and the scientific community insisted on more evidence.

A number of laboratories around the world have, in fact, produced more evidence of the impact. The number of iridium anomalies has increased many fold and includes a wide range of environments, from deep-sea sediments to terrestrial swamp deposits. Furthermore, work by Luck and Turekian (1983) at Yale on osmium isotope ratios has confirmed an extraterrestrial signature in the clays at the end of the Cretaceous. And work by Bohor and his colleagues at the United States Geological Survey on shocked quartz has provided yet more evidence for the impact (Bohor et al. 1984). These and several other lines of argument have made the case for impact very strong, bordering on overkill, with much of the research being stimulated by the controversial nature of extraterrestrial forces as a cause of extinction.

The actual link between impact and extinction is difficult to prove because it is inevitably based on a probabilistic argument. If it could be shown that the extinctions at the end of the Cretaceous occurred *precisely* at the time of the impact, there would probably be no further argument. The difficulty is that, although a number of important biologic groups do in fact disappear at the iridium anomaly, the bulk of the extinctions are not well located in time. It is difficult to say for sure that all the extinctions in the late Cretaceous took place simultaneously and therefore were caused by the impact.

Figure 4.1 illustrates some of the problems with this argument, at least on a local scale. This is a summary of the ranges in time of fifty species of marine brachiopods in one of the better-exposed sections in Denmark from data compiled by Surlyk and Johansen (1984). The twenty species on the left come up to the Cretaceous-Tertiary boundary and disappear rather suddenly. This is certainly compatible with a short-lived event; because the iridium anomaly occurs at this point, the inference of cause and effect is reasonable.

But there is a group of six species (shown by the solid bars) that disappear from the record at about the same time, only to reappear higher in the section. We have no choice but to conclude that these six species did not go extinct. For some reason, they were not preserved through several meters of sediment above the Cretaceous-Tertiary boundary.

The total lack of brachiopod fossils in the first few meters above the boundary leaves open the possibility that the twenty species on the left in figure 4.1 actually died out at scattered times during the barren interval, so that one cannot be certain that there were simultaneous extinctions at all. Alternatively, the barren interval may be seen as a quite reasonable consequence of a severe environmental shock—one that killed many species outright and left the few survivors so depauperate that they left no fossil record. This case, in microcosm, is the kind of problem encountered at all scales.

There are some encouraging developments elsewhere in the extinction record. Several other extinction-iridium pairs have been reported in the literature. The most convincing one, at the Eocene-

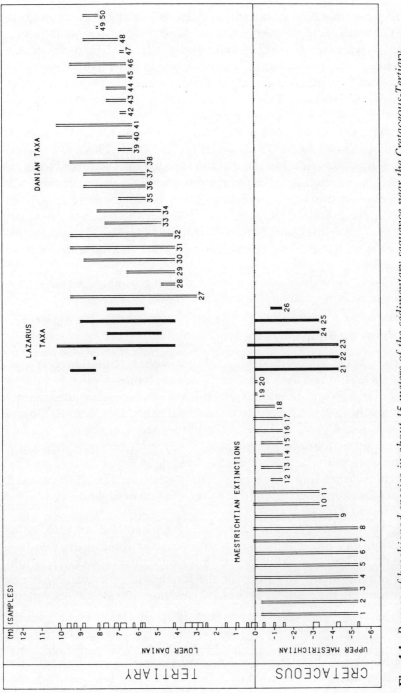

Fig. 4.1. *Ranges of brachiopod species in about 15 meters of the sedimentary sequence near the Cretaceous-Tertiary boundary at Nye Kløv, Denmark. Sample points are shown inside the left scale.*

115

Oligocene boundary, 38 million years before the present, is quite good, with excellent supporting evidence of extraterrestrial material (Alvarez et al. 1982; Ganapathy 1982). But the other events present major problems. The one we would most like to be able to confirm is that at the end of the Permian, the biggest extinction of them all (Sun et al. 1984). The iridium report there, however, has not been replicated in other laboratories.

Two other cases, one in the Jurassic (Brachwicz-Lewinski et al. 1985) and one in the Devonian (Playford et al. 1984), show good iridium anomalies, but the iridium is found only in fossil bacterial mats (stromatolites). The interesting question is whether these are cases of natural, biological enrichment, having nothing to do with high levels of ambient iridium, or whether the bacteria that form the stromatolites concentrated iridium because of a sudden increase in iridium in the environment due to comet or asteroid impact.

We will have to await more data before making a really confident statement about the causes of mass extinction.

ARE EXTINCTIONS PERIODIC?

Whether dealing with stock market averages, climatic records, or skirt lengths, there is a recurring tendency for people to see cycles: a regular ebb and flow of events with repeating sequences. Sometimes the cycles turn out to be real and explicable through simple processes. A good example is the phenomenon of ocean tides. Sea level fluctuates in very regular and predictable ways on daily, semidaily, monthly, and other cycles. In spite of departures caused by local weather conditions, as well as systematic differences in tidal patterns from one region to another, the fact of regular tidal cycles can not be questioned.

Many other cycles in nature have been claimed but turn out to be more in the imagination of the observer than in the realm of reality. Thus, the search for cycles is a quagmire of uncertainty and disappointment. The study of the Earth and of its history is not immune to the quest for regularity, and many proposals of cyclicity have been made. Several of these have involved the spacing of extinctions in time. One such proposal has become a major element in current extinction controversies. In 1984 Jack Sepkoski and I published a short paper arguing on statistical grounds that the major pulses of extinction over the past 250 million years are regularly spaced, occurring every 26 million years (Raup and Sepkoski

1984). Because one or more of the extinction events also bears evidence for Alvarez-style, large-body impact, it was natural to wonder whether impacts of comets or asteroids come in pulses every 26 million years. The Cretaceous mass extinction is one such event, and three of the other possible iridium-extinction pairs are in the pattern. To make matters yet more interesting, three research groups reported that the ages of the larger impact craters on Earth show a compatible periodicity, and several papers have been published claiming the same regularity for reversals in the Earth's magnetic field—a phenomenon which may or may not be caused by comet or asteroid impact (see Raup 1986 for review).

Given these apparent cyclic patterns, it was not surprising that astronomers joined in with hypotheses for cosmic causes of the regularity in timing. Among others, we have the Nemesis theory, which postulates that orbits of comets in our solar system are regularly perturbed by the gravitational effects of an unseen star—a companion to our sun, called Nemesis (Davis et al. 1984).

These developments have given rise to controversy even more heated than that produced by the original impact-extinction hypothesis. A number of good astronomers have published studies arguing that the astronomical explanations of periodic extinction, such as the Nemesis theory, are completely implausible and therefore should not be taken seriously (see Kerr 1985 for review). Other astronomers, however, have been sufficiently convinced to mount searches for the companion star or for other observational evidence for or against the suggested explanations (Muller 1985).

More fundamental is the question of whether extinctions, impact craters, and magnetic reversals are really regularly spaced in time. This gives rise to some difficult problems in the statistical analyses of time series, an area fraught with problems and pitfalls. It is no wonder, therefore, that good statisticians disagree on the validity of the basic claims of periodicity. When it is all over, we may know whether there are large-scale cycles in our solar system or galactic environment that have strongly influenced the history of the Earth and of life—or we may have found one more example of scientists seeing cycles where none exist.

COMMENT

I think the case for a causal link between impact and extinction is much stronger than it was several years ago, but it is still insufficient to satisfy many people. The case for regularly periodicity in the timing of extinctions is even less certain, but the intellectual rewards if the periodicity hypothesis is correct are considerably greater than for the simpler idea of impact causing one or more mass extinctions.

In this context, let me close with a comment about the history of science. The reaction to the idea of impact among rank-and-file paleontologists and geologists has been very negative. Some of the objections are valid because not all parts of the story are equally strong. But I think it is only fair to suggest that some of the negative reactions to the new ideas come from a revival of the old debate between Charles Lyell's uniformitarianism and George Cuvier's catastrophism. Nearly all geologists and paleontologists have been brought up on a strong paradigm which says that to call on extraordinary mechanisms, such as extraterrestrial forces, is unnecessary and to be avoided wherever possible. In his *Principles of Geology* (1833), Lyell asserted: "We are not authorized in the infancy of our science to recur to extraordinary agents. We shall adhere to this plan because history informs us that this method has always put geologists on the road that leads to truth."

This is a very strong statement and it is still heard from graduate students who have never read Lyell. The same admonition to avoid extraordinary explanations was echoed in the now-infamous *New York Times* editorial of April 2, 1985, which ended with the observation that "astronomers should leave to astrologers the task of seeking the causes of earthly events in the stars."

References

Alvarez, L. W., W. Alvarez, F. Asaro, and H. V. Michel. 1980. Extraterrestrial cause for the Cretaceous-Tertiary extinction. *Science* 208: 1,095–1,108.

————. 1982. Iridium anomaly approximately synchronous with terminal Eocene extinctions. *Science* 216:886–88.

Bohor, B. F., E. E. Foord, P. J. Madreski, and D. Triplehorn. 1984. Mineralogic evidence for an impact event at the Cretaceous-Tertiary boundary. *Science* 224:867–69.

Brochwicz-Lewiński, W. A. Gąsiewicz, W. E. Krumbein, G. Melendez, L. Sequeiros, S. Suffczyński, K. Szatkowski, R. Tartowski, and M. Żbik. 1985. Anomalia irydowa nà granicy jury środkowej i górnej. *Przeglad Geologiczny* 32:83–88.

Cuppy, Will. 1983. *How to become extinct*. Reprint. Chicago: University of Chicago Press.

Davis, M., P. Hut, and R. A. Muller. 1984. Extinctions of species by periodic comet showers. *Nature* 308:715–17.

Ganapathy, R. 1982. Evidence for a major meteorite impact on the Earth 34 million years ago: implications for Eocene extinctions. *Science* 216:885–86.

Kerr, R. A. 1985. Periodic extinctions and impacts challenged. *Science* 227:1,451–53.

Luck, J. M., and K. K. Turekian. 1983. Osmium-187/Osmium-186 in manganese nodules and the Cretaceous-Tertiary boundary. *Science* 222:613–15.

Lyell, C. 1833. *Principles of geology.* London: Murray.

Muller, R. A. 1985. Evidence for a solar companion. In *The search for extraterrestrial life: recent developments.* Ed. M. D. Papagiannis. International Astronomical Union.

Playford, P. E., D. J. McLaren, C. J. Orth, J. S. Gilmore, and W. D. Goodfellow. 1984. Iridium anomaly in the Upper Devonian of the Canning Basin, Western Australia. *Science* 226:437–39.

Raup, D. M. 1979. Size of the Permo-Triassic bottleneck and its evolutionary implications. *Science* 206:217–18.

———. 1986. Biological extinction in Earth history. *Science* 231:1,528–33.

Raup, D. M., and J. J. Sepkoski, Jr. 1984. Periodicity of extinctions in the geologic past. *Proceedings of the National Academy of Sciences* 81:801–05.

Shoemaker, E. M. 1984. Large body impacts through geologic time. In *Patterns of change in earth evolution.* Ed. H. D. Holland and A. F. Trendall. Berlin: Springer.

Surlyk, F., and M. B. Johansen. 1984. End-Cretaceous brachiopod extinctions in the Chalk of Denmark. *Science* 223:1,174–77.

Sun, Y.-Y., Z. Chai, S. Ma, X. Mao, D. Xu, Q. Zhang, Z. Yang, J. Sheng, C. Chen, L. Rui, X. Liang, J. Zhao, and J. He. 1984. The discovery of iridium anomaly in the Permian-Triassic boundary clay in Changxing, Zhejiang, China and its significance. In *Developments in geoscience: Contribution to 27th International Geological Congress, Moscow, 1984.* Beijing: Academia Sinica.

Van Valen, L. 1973. A new evolutionary law. *Evolutionary Theory* 1:1–30.

Valentine, J. W., T. C. Foin, and D. Peart. 1978. A provincial model of Phanerozoic marine diversity. *Paleobiology* 4:55–66.

Whitmire, D. P., and A. Jackson IV. 1984. Are periodic mass extinctions driven by a distant solar companion? *Nature* 308:713–15.